T0191955

3D TCAD Simulation for CMOS Nanoeletronic Devices

Yung-Chun Wu · Yi-Ruei Jhan

3D TCAD Simulation
for CMOS Nanoeletronic
Devices

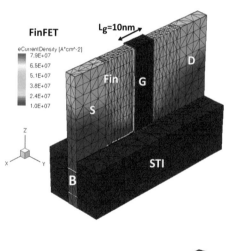

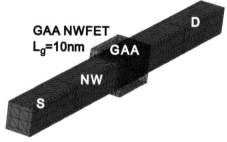

Yung-Chun Wu
Department of Engineering and System
  Science
National Tsing Hua University
Hsinchu
Taiwan

Yi-Ruei Jhan
Department of Engineering and System
  Science
National Tsing Hua University
Hsinchu
Taiwan

ISBN 978-981-10-9779-9   ISBN 978-981-10-3066-6   (eBook)
DOI 10.1007/978-981-10-3066-6

Printed on acid-free paper

This Springer imprint is published by Springer Nature
The registered company is Springer Nature Singapore Pte Ltd.
The registered company address is: 152 Beach Road, #21-01/04 Gateway East, Singapore 189721, Singapore

Preface

Almost on a daily basis, nanoeletronic metal-oxide-semiconductor (CMOS) technology and device design are introduced and explored in rapidly developing semiconductor industry. This book "3D TCAD Simulation for CMOS Nanoeletronic Devices" presents a self-contained and up-to-date critical ideas and illustrations that will help the readers to understand nano electronics device design and its background fundamental physics in detail. Along with basic concepts, the book includes numerous examples which will assist the readers to clearly understand advanced semiconductor research as well. This book will be a proper resource for graduate students doing research in CMOS Nanoeletronic Devices and also for the professional engineers working in both academia and industry. It can also serve as a reference for device research and development engineers and experts in semiconductor industry.

This book reflects the belief that in semiconductor device physics by means of illustrative problems with step-by-step TCAD solutions. This book contents are based on the **Synopsys Sentaurus TCAD 2014 version**. This book thoroughly describes the tools and models for modern nanoeletronic devices by computer simulation technology with which one shall design, develop, and optimize semiconductor device structure and process technology with respect to different important commercialized semiconductor devices and materials. By using TCAD simulation for the analysis of electric and physical properties, time consumed in expensive device fabrication can be minimized leading to effective research output and huge amount of resources and manpower could also be saved. Synopsys Sentaurus TCAD is the leader in global development of 3D TCAD Simulation for CMOS Nanoeletronic Devices. Power houses in semiconductor industry such as Intel, TSMC, Samsung, and IBM are all using the Synopsys products.

This book also considers all the basic semiconductor device physics theory along with recent advanced quantum perspective for nanoelectronic semiconductor device design. It is suggested that readers should have preliminary semiconductor knowledge before reading this book for a better understanding. This book is focused on three main subjects. Part I (Chapters 1–4) are about simulation of electrical and physical properties of Silicon CMOSFET. It starts with the designs of

2D Metal-Oxide-Semiconductor Field-Effect Transistor (MOSFET) and 3D Silicon and Germanium (Lg = 15 and 10 nm) and InGaAs FinFETs. Part II (Chapters 5–7) are about novel nano-semiconductor devices such as Junctionless FET and tunneling FET. Part III (Chapter 8) is about predicting the feasible solutions for Silicon and Germanium FET devices of ultimate minimum dimension and proving that Moore's Law can be extended to the nanotechnology nodes. This chapter on ultra scaled devices serves as only a design guideline and in future more ab-initio and first principle based models shall be incorporated in the device physics for more accurate results which we believe will be updated in future editions of this book.

Instead of direct application of built-in library examples of Synopsys Sentaurus TCAD v. 2014, this book is based on "actual practices of teaching" and "research results" more than 40 international SCI journal papers by our research team in Taiwan National Tsing Hua University over a decade. The design and technology of this book "3D TCAD Simulation for CMOS Nanoeletronic Devices" are fairly important and practical for semiconductor industry and academic research, and it can also improve the development of foresight nanoeletronic semiconductor device. Due to limited knowledge of the author and the continuous update and development of Synopsys Sentaurus TCAD version, users are welcome to contact us via the email address of ycwu.tcad.tw@gmail.com with respect to any mistake or typing errors, or advised to refer to latest user manual of Synopsys Sentaurus TCAD.

Reader can download basic examples at our lab's website http:// semiconductorlab.iwopop.com/. The files are compressed as a zip format. Users should transfer to Synopsys Sentaurus TCAD Workbench under UNIX or Linux system and unzip as directories. Above examples are completely ready to run. Other examples in this book, readers can easily create from above basic examples. We tried to present all the details in a clear and concise method. Thus, readers should be able to follow the computations of all the problems in this book.

We would like to express our deep gratitude to the assistance provided by the research team members of our laboratory in writing this book and the valuable suggestions by students participating in this course over the years. We appreciate Synopsys Company technical support. Also, we would like to acknowledge the Ministry of Science and Technology (MOST) of Taiwan for continuously support, National Nano Device Laboratories (NDL) of Taiwan is greatly appreciated for its technical support in real nanoeletronic devices fabrication. National High-Performance Computing (NCHC) Center of Taiwan is also greatly appreciated for its TCAD simulation support.

Hsinchu, Taiwan   Yung-Chun Wu
2017   Yi-Ruei Jhan

About the book (Modify by author Yung-Chun Wu)

This book demonstrates how to use the Synopsys Sentaurus TCAD 2014 version for the design and simulation of 3D CMOS (complementary metal–oxide–semiconductor) semiconductor nanoelectronic devices, while also providing selected source codes (Technology Computer-Aided Design, TCAD). Instead of the built-in examples of Sentaurus TCAD 2014, the practical cases presented here, based on years of teaching and research experience, are used to interpret and analyze simulation results of the physical and electrical properties of designed 3D CMOSFET (metal–oxide–semiconductor field-effect transistor) nanoelectronic devices, **including Si, Ge, InGaAs FinFET, GAA NWFET, junctionless FinFET, tunnel FinFET. In final chapter, also predicts the feasible options for silicon and germanium FET of ultimate minimum dimensions**.

The book also addresses in detail the fundamental theory of advanced semiconductor device design for the further simulation and analysis of electric and physical properties of semiconductor devices. The design and simulation technologies for nano-semiconductor devices explored here are more practical in nature and representative of the semiconductor industry, and as such can promote the development of pioneering semiconductor devices, semiconductor device physics, and more practically-oriented approaches to teaching and learning semiconductor engineering.

The book can be used for graduate and senior undergraduate students alike, while also offering a reference guide for engineers and experts in the semiconductor industry. Readers are expected to have some preliminary knowledge of the field.

Contents

About the Authors

Dr. Yung-Chun Wu received his B.S. degree in Physics from National Central University in 1996, his M.S. degree in Physics from National Taiwan University in 1998, and his Ph.D. from National Chiao Tung University, Taiwan, in 2005. From 1998 to 2002, he was an assistant researcher at National Nano Device Laboratories, Hsinchu, Taiwan, where he was primarily engaged in research on single electron transistor and electron beam lithography technology. In 2006, he joined the Department of Engineering and System Science, National Tsing-Hua University, Hsinchu, Taiwan, where he is currently working as an associate professor. He teaches 3D CMOS semiconductor nanoelectronic devices by TCAD simulation course for ten years. His research interests include nanoelectronic devices and 3D TCAD simulation, flash memory devices, and solar cells. He has published 56 international SCI papers on nanoelectronic devices.

Yi-Ruei Jhan received the B.S. degree in Physics from National Dong Hwa University in 2010, M.S. degree in Engineering and System Science from National Tsing Hua University in 2012, and Ph.D. degree in Engineering and System Science from National Tsing Hua University in 2015. In 2016, he joined the Research and Development department of Taiwan Semiconductor Manufacturing Company (TSMC) after his graduation. His research interests include Nanoelectronic MOSFET devices, TCAD simulation and Nonvolatile memory devices. He is author of book: 3D TCAD Simulation for CMOS Nanoeletronic Devices.

Chapter 1
Introduction of Synopsys Sentaurus TCAD Simulation

1.1 Introduction

Technology computer-aided design (TCAD) refers to the use of computer simulations to develop and optimize semiconductor processing technologies and devices. **Synopsys Sentaurus TCAD** [1] offers a comprehensive suite of products that includes industry leading process and device simulation tools, as well as a powerful GUI-driven simulation environment for managing simulation tasks and analyzing simulation results. Synopsys Sentaurus TCAD process and device simulation tools support a broad range of applications such as Complementary metal-oxide-semiconductor field-effect transistor **(CMOSFET)**, Fin-shaped field-effect transistor **(FinFET), power devices, memory devices, image sensor, solar cells, and analog/RF devices**. In addition, Synopsys TCAD provides tools for interconnect modeling and extraction, providing critical parasitic information for optimizing chip performance. Synopsys Sentaurus TCAD is dominant simulation software for analysis different semiconductor devices in development and optimization of semiconductor devices in electrical properties, physical properties, and the processing technology simulation. This TCAD tools can replace or partially replace the time-consuming and expensive semiconductor device early research and development. Synopsys Sentaurus TCAD has strong graphical user interface (GUI) visual simulation interface for analysis of simulation result. Synopsys Sentaurus TCAD also provides interaction mode and the extraction tools of **physical properties** and **electric properties** of devices, together with important parameter information of semiconductor device performance, which bring the valuable solution for semiconductor company research and development organizations, and academic organizations.

Synopsys Sentaurus TCAD can be used to predict the important physical and current–voltage properties of current 3D semiconductor devices such as Fin-shaped field-effect transistor (FinFET), and these physical properties include electrical field, electrical potential, electron density, etc.; and electrical properties such as ON

© Springer Nature Singapore Pte Ltd. 2018
Y.-C. Wu and Y.-R. Jhan, *3D TCAD Simulation for CMOS Nanoeletronic Devices*,
DOI 10.1007/978-981-10-3066-6_1

current (I_{on}), OFF current (I_{off}), operation voltage, threshold voltage (V_{th}), operation frequency (f), inverter, and SRAM circuits. All semiconductor processes, device parameters, and impacts on device properties can be analyzed by the design of 3D semiconductor devices with different structures and materials. We can analyze the strengths and weaknesses of 3D device's key performance indicator (KPI). This TCAD tool can greatly reduce the time and cost for R&D of top semiconductor companies (such as TSMC, Intel, Samsung, IBM, UMC, Global foundry).

The purpose of this book is to not only allow readers to understand and use the **Synopsys Sentaurus TCAD 2014 version** for design and simulation of 3D semiconductor device based on integrated fundamental theory of semiconductor device, but also simulate the electrical and physical properties of advanced 3D semiconductor device in conjunction with the capability of software-aided design of 3D semiconductor devices.

This book emphasizes three major subjects: Part I (Chaps. 1–4) are about the simulation of electrical properties of silicon CMOSFET, starting with the designs of 2D MOSFET and 3D silicon FinFET CMOS devices and circuits; Part II (Chaps. 5–7) are about advanced nanoscale semiconductor devices such as Chap. 5: GAA NWFET, Chap. 6: junctionless FET and Chap. 7: tunneling FET; Part III (Chap. 8) is about predicting the feasible options for silicon and germanium FET of ultimate minimum dimensions.

Instead of the built-in examples of Sentaurus TCAD 2014, the examples in this book are the practical cases in years of our group research results and teaching course in National Tsing Hua University, Hsinchu, Taiwan. The design and simulation technologies of nano-semiconductor device discussed in this book are rather practical and representative in semiconductor industry and among academic semiconductor researches. This book can help the development of advanced nanoscale semiconductor device; understand semiconductor device physics, and the practically learn the semiconductor engineering by TCAD simulation. This book is suitable for the learning by graduated students who engaged COMS nanoeletronic devices. It also can serve as the reference for engineers and experts in semiconductor industry.

1.2 Introduction of Moore's Law and FinFET

Moore's law was proposed by Gordon Moore, one of the founders of Intel. It is the observation that the number of transistors in a dense integrated circuit doubles approximately every 18–24 months. This trend has continued for over half a century. Moore's law is actually a prediction of development of semiconductor industry rather than a real law of physics. It is expected that Moore's law is expected to hold until 2030. According to 2015 International Technology Roadmap for Semiconductors (ITRS) version 2.0 [2], the device miniaturization development over the past few years is as shown in Fig. 1.1, and the R&D process of logic device of Intel [3] is as shown in Fig. 1.2.

YEAR OF PRODUCTION	2015	2017	2019	2021	2024	2027	2030
Logic device technology naming	P70M56	P48M36	P42M24	P32M20	P24M12G1	P24M12G2	P24M12G3
Logic industry "Node Range" Labeling (nm)	"16/14"	"11/10"	"8/7"	"6/5"	"4/3"	"3/2.5"	"2/1.5"
Logic device structure options	FinFET FDSOI	FinFET FDSOI	FinFET LGAA	FinFET LGAA VGAA	VGAA,M3D	VGAA,M3D	VGAA,M3D
LOGIC DEVICE GROUND RULES							
MPU/SoC Metalx $^1/_2$ Pitch (nm)	28.0	18.0	12.0	10.0	6.0	6.0	6.0
MPU/SoC Metal0/1 $^1/_2$ Pitch (nm)	28.0	18.0	12.0	10.0	6.0	6.0	6.0
L_g : Physical Gate Length for HP Logic (nm)	24	18	14	10	10	10	10
L_g : Physical Gate Length for LP Logic (nm)	26	20	16	12	12	12	12
FinFET Fin Width (nm)	8.0	6.0	6.0	NA	N/A	N/A	N/A
FinFET Fin Height (nm)	42.0	42.0	42.0	NA	N/A	N/A	N/A
Device effective width - [nm]	92.0	90.0	56.5	56.5	56.5	56.5	56.5
Device lateral half pitch (nm)	21.0	18.0	12.0	10.0	6.0	6.0	6.0
Device width or diameter (nm)	8.0	6.0	6.0	6.0	5.0	5.0	5.0
DEVICE PHYSICAL&ELECTRICAL SPECS							
Power Supply Voltage - V_{dd} (V)	0.80	0.75	0.70	0.65	0.55	0.45	0.40
Subthreshold slope - [mV/dec]	75	70	68	65	40	25	25
Inversion layer thickness - [nm]	1.10	1.00	0.90	0.85	0.80	0.80	0.80
$V_{t,sat}$ (mV) at I_{off}=100nA/um - HP Logic	129	129	133	136	84	52	52
$V_{t,sat}$ (mV) at I_{off}=100pA/um - LP Logic	351	336	333	326	201	125	125
Effective mobility (cm2/V.s)	200	150	120	100	100	100	100
Rext (Ohms.um) - HP Logic [7]	280	238	202	172	146	124	106
Ballisticity.Injection velocity (cm/s)	1.20E-07	1.32E-07	1.45E-07	1.60E-07	1.76E-07	1.93E-07	2.13E-07
V_{dsat} (V) - HP Logic	0.115	0.127	0.136	0.128	0.141	0.155	0.170
V_{dsat} (V) - LP Logic	0.125	0.141	0.155	0.153	0.169	0.186	0.204
Ion (uA/um) at I_{off}=100nA/um - HP logic w/ Rext=0	2311	2541	2782	2917	3001	2670	2408

Fig. 1.1 Selected logic core device technology road map as predicted by 2015 ITRS version 2.0 [2]

The industry also is working on sub-7-nm-technology node by year 2020. Unfortunately, 5-nm technology presents a multitude of unknowns and challenges. For one thing, the exact timing and specs of 5 nm remain unclear. Then, there are several technical and economic roadblocks. And even if 5 nm happens, it is likely that only a few companies will be able to afford it. By now, a great deal of resources has been dedicated by the semiconductor sector into the scaling of size of CMOS device for extending Moore's law to the sub-7-nm CMOS technology. Many challenge will appear when the feature size of device is approaching sub-7-nm-technology node, such as the new device structure, new material issue, short-channel effect (SCE), and power consumption. For now, gate-all-around (GAA) is generating the most possibility, although this technology presents several challenges in the fab. Making the patterns, gates, nanowires, and interconnects are all challenging. In addition, process control could be a remarkable challenge. And, of course, the ability to make gate-all-around field-effect transistor (GAA FET) in a cost-effective manner is key issue (Figs. 1.3 and 1.4).

In addition, 2015 ITRS [2] also predicted that tri-gate monolithic 3D (M3D) or vertical GAA FET may a solution in sub-7-nm-semiconductor technology node. Figure 1.1 shows the important device design parameters.

In addition, semiconductor manufacturing companies first decide on the channel materials for the pFET and nFET structures. The options for pFET are silicon, germanium (Ge), or SiGe. For the nFET, silicon, SiGe, Ge, or an III–V material could be used.

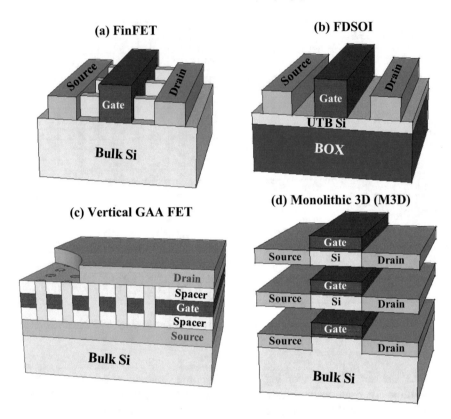

Fig. 1.2 Schematic plots of **a** FinFET, **b** fully depleted silicon-on-insulator FET, **c** vertical nanowire gate-all-around FET, and **d** monolithic 3D FET, after 2015 ITRS version 2.0

With the world's attention, Intel Developer Forum (IDF) [3] was held in San Francisco, USA, 2015. Intel indicated in this forum that Moore's law would continue to lead the breadth and speed of "innovation and integration" based on the company's technical advantages of nanoscale processes. In **2015, Intel introduced the new-generation 14-nm Fin-shaped field-effect transistor (FinFET) CPU Broadwell platform** by adopting the advanced fabrication process of 14-nm FinFET CPU together with the Intel second-generation 3D FinFET technology. Intel is the first semiconductor company to enter the 14 nm era, and Broadwell CPU will be the first to adopt this advanced process. The ultra-low voltage Core M series customized for "Y" series CPU for ultra-slim tablet PC has been launched to the market at the end of 2015. A part of details of 14-nm technology was publicly disclosed by Intel in 2014 IDF: The thermal design power (TDP) of the new product is only less than half of the previous generation, while it can provide similar performance with better lifetime. Intel Broadwell structure has been optimized with respect to the advantage of new feature of 14-nm process by adopting the second-generation FinFET. It will be applied to various high-performance

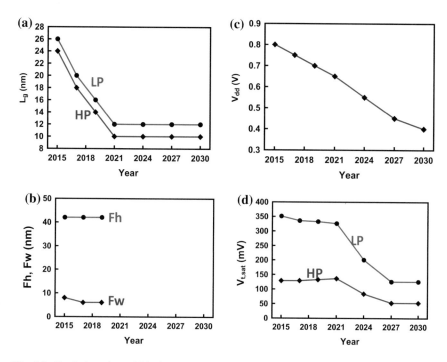

Fig. 1.3 Prediction plots of 2015 ITRS for **a** physical gate length (L_g) for HP and LP, **b** Fh and Fw, **c** V_{dd}, and **d** Vt_{sat} for HP at I_{off} = 100 nA/μm and Vt_{sat} for LP at I_{off} = 100 pA/μm. *HP* high-performance technology and *LP* low-power technology. *Fh* Fin height of FinFET, *Fw* Fin width of FinFET

low-power consumption products such as smartphones, PCs, servers, large workstations, and Internet-of-things (IOT) applications.

From this Fig. 1.5, it appears that the Fin shape of second-generation 14-nm FinFET is taller and narrower, like a wine-bottle shape for improving gate control capability and higher on-state current (I_{on}) (Fig. 1.6).

1.3   Sentaurus Window Environment and Workbench for TCAD Task Management

Synopsys Sentaurus TCAD is a complete graphical operating environment for establishment, management, execution, and analysis of TCAD simulation. The intuitive graphical interface allows users to automatically process and easily operate TCAD simulation with high efficiency, making it an excellent information management solution for semiconductor simulation program. It includes preprocessing of coding documents entered by users, extraction of KPI parameters by simulation

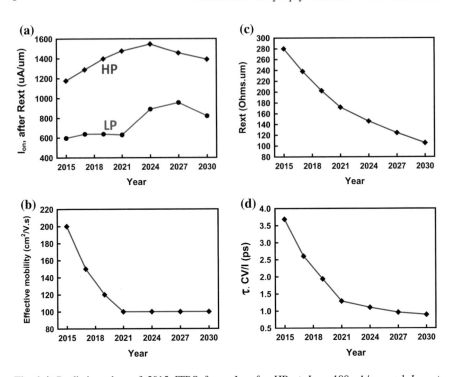

Fig. 1.4 Prediction plots of 2015 ITRS for **a** I_{dsat} for HP at $I_{off} = 100$ nA/μm and I_{dsat} at $I_{off} = 100$ pA/μm, **b** effective mobility, **c** source/drain resistance, and **d** intrinsic delay (CV/I)

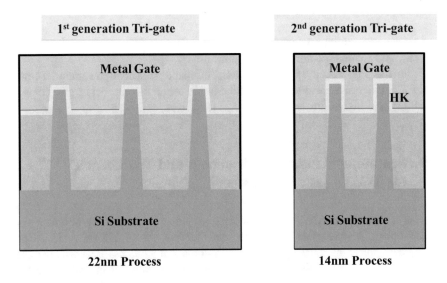

Fig. 1.5 Differences of shapes between 2015 Intel first-generation high-k metal gate (*HKMG*) FinFET (or called tri-gate FET) and second-generation HKMG FinFET

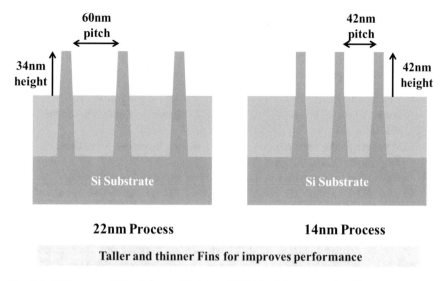

Fig. 1.6 Differences of Fin pitches and heights of 2015 Intel first-generation 22-nm FinFET and second-generation 14-nm FinFET

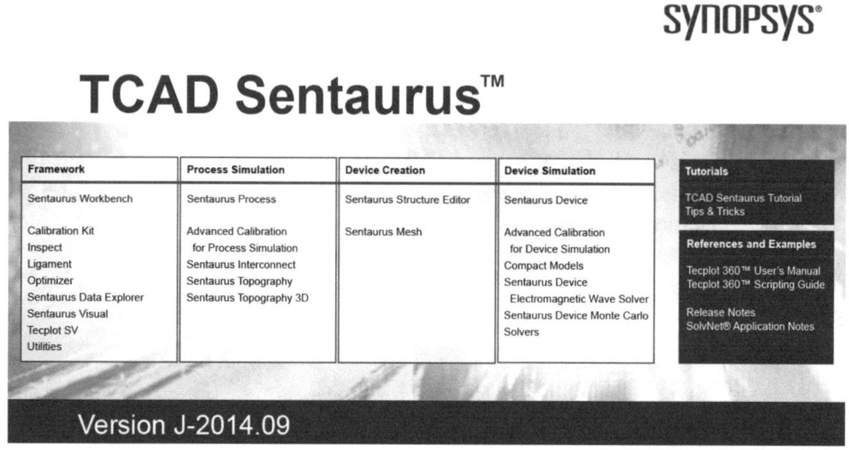

Fig. 1.7 Synopsys Sentaurus TCAD is a complete graphical operating environment which includes numerous simulation tools (Copyright © Synopsys, Inc. All rights reserved.)

tools, setting of important variables, and planning process flow for a project. The simulation results can be presented in the form of visual display. The simulation raw data can also be exported via proper graphical analysis software for analyzing electrical and physical properties (Fig. 1.7).

1.4   Synopsys Sentaurus TCAD Software and Working Environment

Features of Synopsys Sentaurus TCAD

(1) High efficiency and streamlined management of simulation items.
(2) Automatic processing and simplification of large-scale simulation via minimum user interaction.
(3) Convenient folder hierarchical representation of technical simulation.
(4) Fully parametric simulation.
(5) Optimization and sensitivity analysis which are easy to implement.
(6) Precise 1D, 2D, and 3D visual displays of TCAD structures and simulation results (Fig. 1.8).

Sentaurus device is used to simulate the electrical characteristics of the device.

Finally, Sentaurus Visual is used to visualize the output from the simulation in 2D and 3D, and inspects used to plot the electrical characteristics (Fig. 1.9).

The basic process flowchart of semiconductor device simulation by Synopsys Sentaurus TCAD 2014 version and the required simulation tools in this book are shown in Fig. 1.10.

(1) Sentaurus Workbench (SWB)

SWB includes a toolbar and a graphical interface for establishing, editing, and organizing technical process flow. The higher level architecture supports user-defined database, which can reflect the processes and results of semiconductor fabrication process technology or electrical property tests. User can use Sentaurus Workbench to automatically generate experimental design groups and to allocate simulation operations in computer network.

Synopsys Sentaurus TCAD user interface is shown in Fig. 1.11.

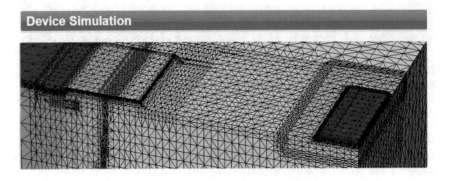

Fig. 1.8 Tools for simulation device performance

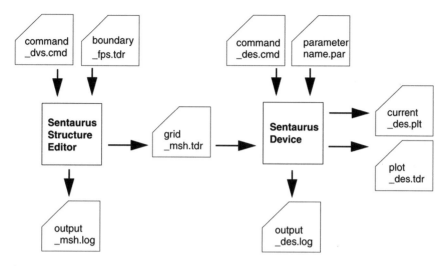

Fig. 1.9   Typical tool flow with device simulation using Sentaurus Device

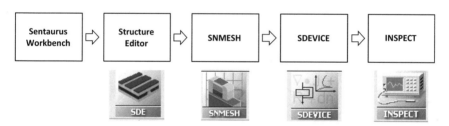

Fig. 1.10   Basic process flow charts of simulation tools by Synopsys Sentaurus TCAD 2014 version and the required simulation tool software

(2) **Sentaurus Device Editor (SDE)**

Structure Editor is a tool for creating device geometric structure of TCAD simulation. **Structure Editor** is an editor combining 2D and 3D device geometric structures, and it is also a simulation tool developed by the TCAD-based 3D technology. There are different operating modes integrated in this editor, all of which share the same data representation. The drawing of geometric structure and the 3D device geometric structure established by syntax can be freely mixed and matched to generate any 3D structure with great flexibility. In addition, **Structure Editor** provides the most advanced visualization technology. The structure can be timely examined during establishment process. This powerful visualization software allows users to select certain area to be displayed while leaving other areas transparent or not displayed, thus effectively improving the design efficiency of developers.

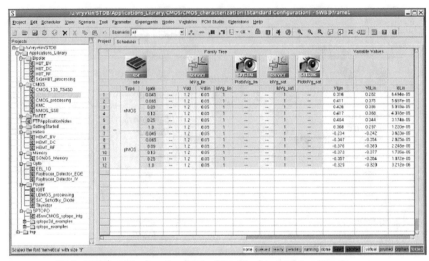

Project Editor main view showing traditional horizontal flow orientation

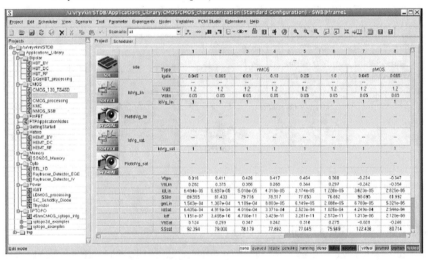

Fig. 1.11 Synopsys Sentaurus TCAD user interface

Features of Structure Editor:

1. Establishing 2D and 3D structures by direct TCAD operation and technical simulation steps.
2. User-friendly interactive user interface and the most advanced visualization technology.
3. Graphical user interfaces and mesh engine.

4. Descriptive command bar which can be accessed and recorded from graphical interface.

In this book, we start from **SDE** tool to establish **3D nanoelectronic device** structure by defining several "blocks." With given device's dimension, materials, and different dopant of each block, complicated structures such as FinFET or GAA FET can be easily created by arrangement and combination of these blocks. SDE tool also allows definition of variable parameters for subsequent adjustment on **SWB**, such as thickness of gate oxide, length of gate, metal work function, and operating voltages. Other important semiconductor technologies, such as silicide, high dielectric materials, metal gates, lightly doped drain (LDD), and body bias, can also be easily designed and simulated in Synopsys Sentaurus TCAD.

For example, the FinFET structure based on silicon bulk is shown in Fig. 1.12.

(3) **SNMESH**

SNMESH tool refers to the points of mathematic model to be solved, where the density of mesh can be self-defined. The location with denser mesh can better reflect the variation of physical properties of this area, such as potential gradient, electric field gradient, and carrier concentration gradient. Excessive mesh will result in prolonged simulation time.

For example, the mesh on FIN structure of Bulk FinFET is shown in Figs. 1.13 and 1.14.

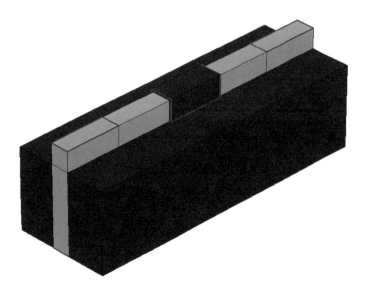

Fig. 1.12 FinFET structure on bulk is established by the permutation and combination of 3D blocks

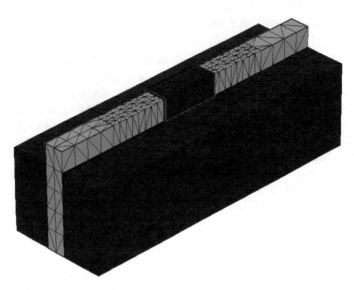

Fig. 1.13 Mesh for TCAD simulation of bulk FinFET

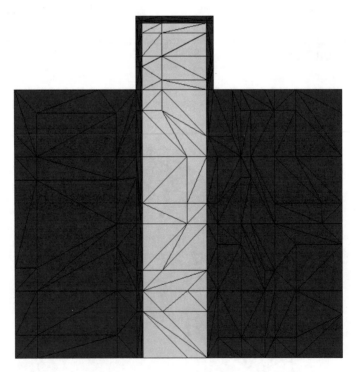

Fig. 1.14 2D cross-sectional view of mesh on Fin structure of bulk FinFET

(4) SDEVICE

SDEVICE tool is a general-purpose device simulation tool which offers simulation capability in the following broad categories:

(a) Advanced Logic Technologies: Sentaurus Device simulates advanced logic technologies such as Si FinFET and FDSOI, including stress engineering, channel quantization effects, hot carrier effects and ballistic transport, and many other advanced transport phenomena. Sentaurus Device also supports the modeling of SiGe, SiSn, InGaAs, InSb, and other high-mobility channel materials and implements highly efficient methods for modeling atomistic and process variability effects.

(b) Compound Semiconductor Technologies: Sentaurus Device can simulate advanced quantization models including rigorous Schrödinger solution and complex tunneling mechanisms for transport of carriers in heterostructure devices such as HEMTs and HBTs made from, but not limited to, GaAs, InP, GaN, SiGe, SiC, AlGaAs, InGaAs, AlGaN, and InGaN.

(c) Optoelectronic Devices: Sentaurus Device has the capability to simulate the optoelectronic characteristics of semiconductor devices such as CMOS image sensors and solar cells. Options within Sentaurus Device also allow for rigorous solution of the Maxwell's wave equation using FDTD methods.

(d) Power Electronic Devices: Sentaurus Device is the most flexible and advanced platform for simulating electrical and thermal effects in a wide range of power devices such as IGBT, power MOS, LDMOS, thyristors, and high-frequency high-power devices made from wide bandgap material such as GaN and SiC.

(e) Memory Devices: With advanced carrier tunneling models for gate leakage and trapping de-trapping models, Sentaurus Device can simulate any floating gate device like SONOS and flash memory devices including devices using high-k dielectric.

(f) Novel Semiconductor Technologies: Advanced physics and the ability to add user-defined models in Sentaurus Device allow for investigation of novel structures made from new material.

(5) INSPECT

INSPECT tool is used for extracting current and voltage properties of semiconductor device, such as:

1. Subthreshold swing (SS).
2. Threshold voltage (V_{th}).
3. Drain-induced barrier lowering (DIBL).
4. Transconductance (G_m).
5. Saturation current (I_{sat}).
6. Off-state leakage current (I_{off}).
7. Resistance (R_{out})

8. Inverter performance
9. SRAM performance
10. Analog/RF performance.

1.5 Simulation Project View on Sentaurus Workbench (SWB)

The **SWB family tree** view of simulation project is as shown in Fig. 1.15 with user-friendly window-based user interface.

Sentaurus TCAD Toolbar Buttons of user-friendly window-based user interface are shown in Fig. 1.16.

1.6 Sentaurus Visual

Sentaurus Visual tool is the advanced visualization software for TCAD data analysis. It is equipped with rich graphics capabilities for interactive composition of X–Y curves and 2D/3D TCAD device structures and device electrical and physical properties. The 2D and 3D user interfaces of Sentaurus Visual are shown in Figs. 1.17 and 1.18.

In addition, semiconductor device technology integrated with virtual process is user-friendly. Semiconductor device simulation technologies such as Front End of Line (FEOL) and Back End of Line (BEOL) can all be processed by tools such as Sentaurus Interconnect. The strong mathematical simulation algorithm is capable of simulating technical steps of ion implantation, thermal diffusion, doping activation,

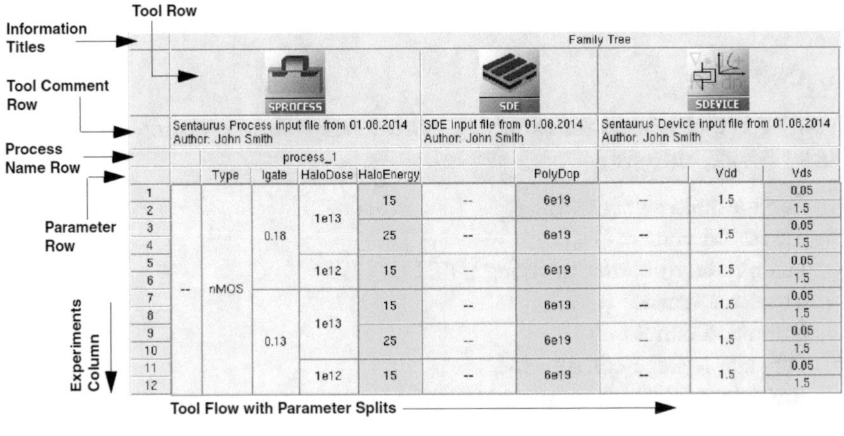

Fig. 1.15 The family tree view of simulation project of SWB

Table 1   Project Editor toolbar buttons

Button	Description	Button	Description
	Creates a new project		Quickly visualizes output files for selected nodes with the default visualizer
	Opens a project		Displays information (properties) about the currently selected item
	Saves a project under an existing name		Runs current project or selected nodes
	Reloads a project		Terminates running project or selected nodes
	Stops the loading of a project		Zooms in to the project view
	Closes the currently opened project		Zooms out of the project view
	Cuts a selection		Resets zoom to the default
	Copies a selection		Changes the current project view to the next view
	Pastes a cut or copied selection		Opens a spreadsheet application with the current view
	Undoes previous operation		Displays a simulation project vertically
	Adds a new tool, parameter, experiment, or variable to the tree		Displays a simulation project horizontally
	Deletes selected tool, parameter, experiment, or variable from the tree		Displays a simulation project in compact mode
	Adds a new experiment to the tree (the default values are preset to a selected experiment if there is one)		Displays a simulation project in full mode
	Adds parameter values to either all (full factorial) or selected experiments		Opens a command prompt in a project directory as a separate shell
	Opens submenu for editing input files of selected tool		Opens manuals in PDF format using the Adobe® Reader® application
	Visualizes output files for selected nodes		Opens TCAD Sentaurus Tutorial (HTML training material) in Web browser

Fig. 1.16 Detailed explanation of function of toolbar buttons

etching, material deposition, oxidation, and epitaxy with respect to different semiconductor materials. The process conditions of gas composition, temperature, and pressure in each technical step are the typical input conditions. The final product is the 2D or 3D device structure to be used for device electrical stimulation. Synopsys is equipped with four technical simulation tools to be selected by users: Sentaurus Process; Taurus TSUPREM-4; Sentaurus Lithography; and Sentaurus Topography.

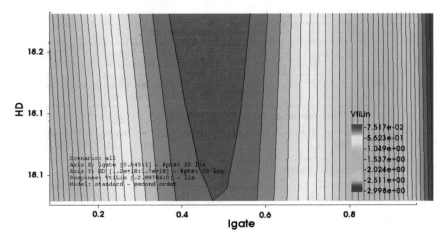

Fig. 1.17 Sentaurus Visual 2D user interface

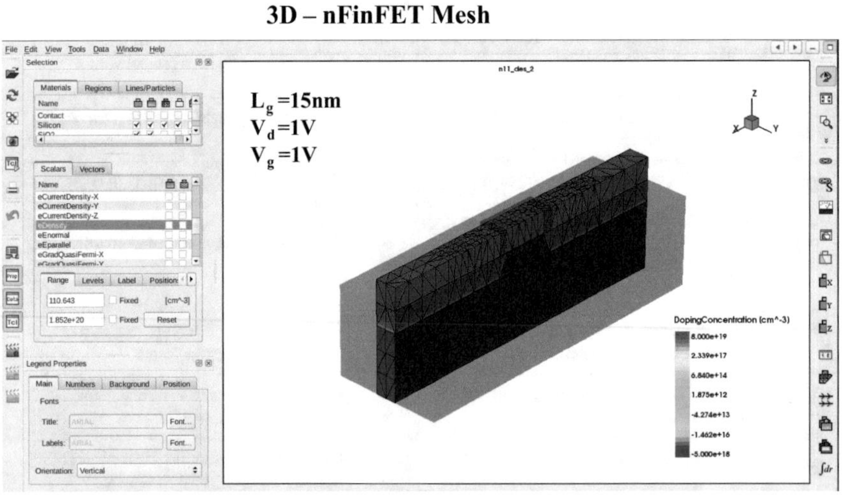

3D – nFinFET Mesh

Fig. 1.18 Sentaurus Visual 3D n-type FinFET user interface

1.7   Calibration and Services

Calibration and consulting services are also available by Synopsys company support. Currently, all forefront technologies and device engineers of semiconductor companies are using Synopsys TCAD tool to develop and optimize forefront semiconductor device technologies. Sometimes, these tools need to be calibrated with respect to specific technology in order to improve the predictability of future nodes or special process. Synopsys TCAD provides customers with calibration, simulation, model development, and integration services. The calibration and services provided by Synopsys based on complete and verified TCAD technology and TCAD device module solutions can accelerate technological development and solve problems of fabrication process of forefront semiconductor companies, thus enhancing the international competitiveness of semiconductor company.

References

1. Synopsys Sentaurus TCAD Ver. J-2014.09, Synopsys, Inc., Mountain View, CA, USA. (2014). https://www.synopsys.com/home.aspx
2. ITRS version 2.0. (2015). http://www.semiconductors.org/main/2015_international_technology_roadmap_for_semiconductors_itrs/
3. Intel Developer Forum on line material, San Francisco, CA, USA. (2014)

Chapter 2
2D MOSFET Simulation

2.1 Complementary MOS (CMOS) Technology

Since the development of metal-oxide-semiconductor field-effect transistor (MOSFET) started in 1950s, relevant technologies have been constantly improving [1]. An N-channel MOSFET is shown in Fig. 2.1a, which is also known as nMOSFET or nFET. It is called N-channel because the conduction channel (which is the inversion layer) is filled with inversion electrons (N-type carriers) as shown in Fig. 2.1b. Figure 2.1c, d both are showing P-channel MOSFET (inversion holes), which is also known as pMOSFET or pFET. The V_g and V_d of these two kinds of transistors are both ranging from 0 V to V_{dd}. The body of nFET is connected to the lowest voltage of circuit, 0 V, as shown in Fig. 2.1b. Therefore, the PN junction is always in a reverse bias or no bias such that there will not be any forward bias current. When V_g is equaled to V_{dd} and larger than threshold voltage (V_{th}) $V_g = V_{dd} > V_{th}$, the electron inversion layer will appear, and the nFET will be conducting. On the other hand, the source and body of pFET are connected to V_{dd} as shown in Fig. 2.1d, thus the applied V_g is just the opposite of nFET. When $V_g = V_{dd}$, the nFET is conducting and the pFET is off. On the contrary, when $V_g = 0$, nFET is off and pFET is conducting.

The low power circuit designed based on the complementary characteristic of nFET and pFET is called complementary MOS (or CMOS) circuit as shown in Fig. 2.2a. There is a circle at the gate with the circuit symbol for pFET, meaning this circuit is an inverter. It will charge the load capacitor C at the output terminal to V_{dd} or discharge it to 0 V in accordance with the command of gate voltage V_g. When $V_g = V_{dd}$, nFET is conducting and pFET is off (these two transistors are regarded as a simple switch), and the output terminal is pulled down to ground point ($V_{out} = 0$). When $V_g = 0$, nFET is off and pFET is conducting, and the output terminal V_{out} is

© Springer Nature Singapore Pte Ltd. 2018
Y.-C. Wu and Y.-R. Jhan, *3D TCAD Simulation for CMOS Nanoeletronic Devices*,
DOI 10.1007/978-981-10-3066-6_2

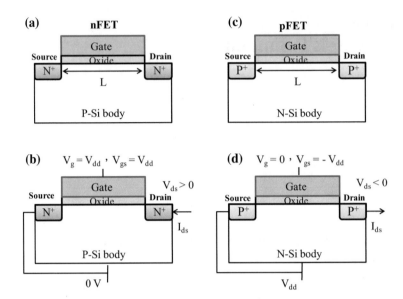

Fig. 2.1 a N-channel FET is in off-state, **b** N-channel FET is in on-state, **c** P-channel FET is in off-state, **d** P-channel FET is in the on-state [1]

pulled up to V_{dd}. Figure 2.2b shows how nFET and pFET are fabricated on the same chip. An N-type well is formed in a portion of area in the P-type silicon substrate by implantation or diffusion of N-type dopant. The contacts of P-type silicon substrate and N-type well are both shown in the figure. The layout of basic CMOS inverter is as shown in Fig. 2.2c. This is an image of top–down view of silicon water, which is also the image composed of several overlapped masks for fabrication of inverter. V_{in}, V_{out}, V_{dd}, and ground point voltage are all based on metal lines. Polysilicon gate or metal gate is the vertical line connected to V_{in}.

In this chapter, the physical and electrical properties of MOSFET will be investigated by 2D TCAD simulation. It is shown in Fig. 2.2 that the current is flowing from the high potential terminal to the low potential terminal.

The current (I_{ds}) in Fig. 2.3 can be derived from the charge concentration and width of inversion layer as shown below:

$$I_{ds} = W \times Q_{inv}(x) \times v = W \times Q_e \times \mu_{ns} \times E$$
$$= W \times C_{oxe}(V_{gs} - V_{cs} - V_t)\mu_{ns}dV_{cs}/dx \tag{2.1}$$

$$\int_0^L I_{ds}dx = WC_{oxe}\mu_{ns} \int_0^{V_{ds}} (V_{gs} - V_{cs} - V_t)dV_{cs} \tag{2.2}$$

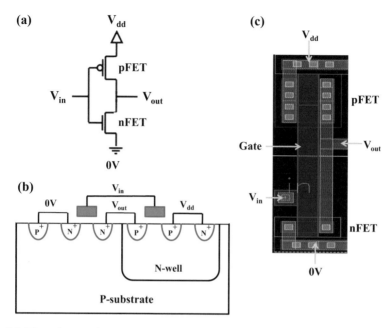

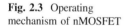

Fig. 2.2 Three figures of CMOS Inverter, **a** CMOS inverter circuit consists of a pFET pull-up device and a nFET pull-down device, **b** nFET and pFET device structures are integrated in a same chip, and **c** the layout of CMOS inverter

Fig. 2.3 Operating mechanism of nMOSFET

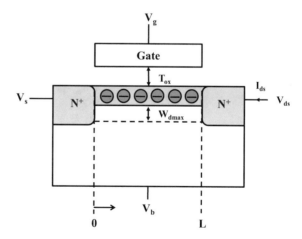

$$I_{ds}L = WC_{oxe}\mu_{ns}\left(V_{gs} - V_t - \frac{1}{2}V_{ds}\right)V_{ds} \tag{2.3}$$

$$\boxed{I_{ds} = \frac{W}{L}C_{oxe}\mu_{ns}\left(V_{gs} - V_t - \frac{1}{2}V_{ds}\right)V_{ds}} \tag{2.4}$$

Equation (2.4) reveals that I_{ds} is proportional to W, μ_{ns}, V_{ds}/L (average electric field in the channel), and $C_{ox}(V_g - V_t - \frac{1}{2}V_{ds})$. The $C_{ox}(V_g - V_t - \frac{1}{2}V_{ds})$ can reflect the inversion electron density Q_{inv} (C/cm^2) in the channel. C_{oxe} is effective gate capacitance (F/cm^2).

When V_{ds} is very small, the item of 1/2 V_{ds} of (2.4) can be neglected, so I_{ds} is proportional to V_{ds}, which means the conduction behavior of transistor under such voltage making it just like a resistor. The I–V When V_{ds} is increased, Q_{inv}, and dI_{ds}/dV_{ds} are decreased. The differentiation of Eq. (2.4) with respect to V_{ds} will result in dI_{ds}/dV_{ds}, which will be equaled to 0 under a specific V_{ds}. At this moment, the V_{ds} is called V_{dsat} as shown below:

$$\frac{dI_{ds}}{dV_{ds}} = 0 = \frac{W}{L}C_{oxe}\mu_{ns}(V_{gs} - V_t - V_{ds}), \quad \text{when} \quad V_{ds} = V_{dsat} \tag{2.5}$$

$$\boxed{V_{dsat} = V_{gs} - V_t} \tag{2.6}$$

V_{dsat} is called drain saturation voltage, and different V_{gs} will lead to different V_{dsat}. The region with V_{ds} far less than V_{dsat} is called linear region, and the region with V_{ds} greater than V_{dsat} is called "saturation region." The part of I–V curve of Fig. 2.4 with $V_{ds} \ll V_{dsat}$ is the "linear region."

Fig. 2.4 Output characteristics of nMOSFET

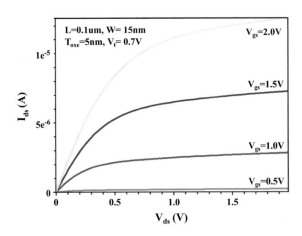

The current in saturation region can be derived from Eq. (2.4) as:

$$\boxed{I_{dsat} = \frac{W}{2L} C_{oxe} \mu_{ns} (V_{gs} - V_t)^2}$$

(2.7)

In addition, the transconductance is defined as:

$$g_m \equiv dI_{ds}/dV_{gs}|_{V_{ds}}$$

(2.8)

And the saturation transconductance is defined as:

$$g_{msat} = \frac{W}{L} C_{oxe} \mu_{ns} (V_{gs} - V_t)$$

(2.9)

The simulation and analysis of 2D MOSFET will be described in the following sections. Designers often refer to this region as the active region. In the last section, we will discuss about TCAD Simulation of 2D nMOSFET and pMOSFET.

2.2 [Example 2.1] 2D n-Type MOSFET with I_d–V_g Characteristics Simulation

The N-type 2D nMOSFET with different L_g = 200, 400, 600, 800, 1000 nm is used as an example for the introduction of simulation technology. First, the Synopsys Sentaurus TCAD 2014 version is used to establish the **four tools of SDE, SNMESH, SDEVICE, and INSPECT** as shown in Fig. 2.5.

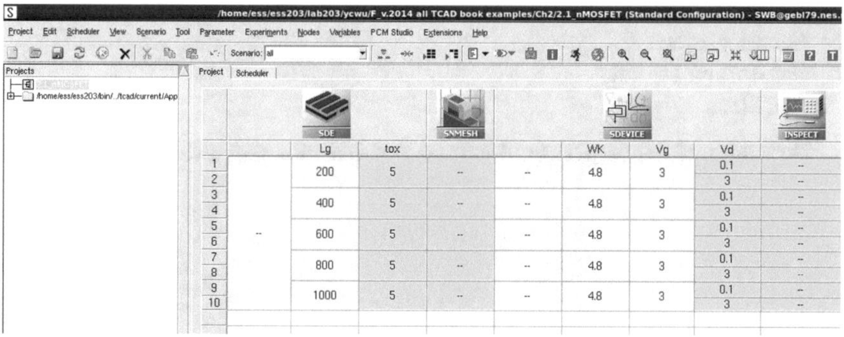

Fig. 2.5 Required simulation four tools are shown in the workbench of 2D MOSFET simulation

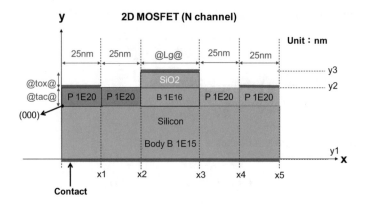

Fig. 2.6 Cartoon plot of 2D MOSFET structure for simulation

Now, we start to establish the device structure. The commands icon on **SDE** tool should be right-clicked to enter codes to establish the structure. At first, we need to draw a cartoon plot as shown in Fig. 2.6 of 2D MOSFET structure for simulation.

The structure of 2D MOSFET to be established is as shown in Fig. 2.6.

1. SDE → devise_dvs.cmd

Now, we use Fig. 2.6 to explain the SDE tool commands (code file is devise_dvs. cmd)

In principle, the flow of structure establishment is as shown below:

(1) **Set zero point and coordinates**
(2) **Composition of 2D Structure**
(3) **Composition of 3D Structure and 2D Y-cut diagram (in** Chap. 3)
(4) **Set 2D rectangles or 3D cuboids (in** Chap. 3)
(5) **Set Electrodes**
(6) **Set doping region**
(7) **Set Mesh**

The codes of devise_dvs.cmd can be divided into six parts.

(1) **Parameter**
(2) **Structure**
(3) **Contact**
(4) **Doping**
(5) **Mesh**
(6) **Save**

The complete codes of this 2D MOSFET example are as shown below:

```
;----- (1). parameter -----;
(define Lg @Lg@)        ;set Lg as variables Lg = 200, 400, 600, 800, 1000 nm
(define tox @tox@)       ;set tox as variables
(define tac 100)
(define Body 400)
(define LSDC 25)
(define LSD 25)
(define C_Doping 1e16)
(define D_Doping 1e20)
(define S_Doping 1e20)
(define B_Doping 1e15)
(define nm 1e-3)

(define x1 LSDC)
(define x2 (+ x1 LSD))
(define x3 (+ x2 Lg))
(define x4 (+ x3 LSD))
(define x5 (+ x4 LSDC))

(define y1 (- Body))
(define y2 tac)
(define y3 (+ tac tox))
;----- (2). Structure -----;
"ABA"
;--- source ---
(sdegeo:create-rectangle
   (position     0 0 0)
   (position     x1 y2 0)   "Silicon" "SourceC" )
(sdegeo:create-rectangle
   (position     x1 0 0)
   (position     x2 y2 0)   "Silicon" "Source" )
;--- Channel ---
(sdegeo:create-rectangle
   (position     x2 0 0)
   (position     x3 y2 0)   "Silicon" "Channel" )
```

```
;--- Drain ---
(sdegeo:create-rectangle
    (position        x3 0 0)
    (position        x4 y2 0)    "Silicon" "Drain" )
(sdegeo:create-rectangle
    (position        x4 0 0)
    (position        x5 y2 0)    "Silicon" "DrainC" )
;--- Body ---
(sdegeo:create-rectangle
    (position        0 0 0)
    (position        x5 y1 0)    "Silicon" "Body" )
;--- Gate oxide ---
(sdegeo:create-rectangle
    (position        x2 y2 0)
    (position        x3 y3 0)    "SiO2" "Gateoxide" )


; ----- (3). Contact -----;
;----- Gate -----
(sdegeo:define-contact-set "G" 4.0    (color:rgb 1.0 0.0 0.0 ) "##")
(sdegeo:define-2d-contact    (find-edge-id (position (+ x2 (/ Lg 2)) y3 0)) "G")
;----- Source -----
(sdegeo:define-contact-set "S" 4.0    (color:rgb 1.0 0.0 0.0 ) "##")
(sdegeo:define-2d-contact    (find-edge-id (position 10 tac 0)) "S")
;----- Drain -----
(sdegeo:define-contact-set "D" 4.0    (color:rgb 1.0 0.0 0.0 ) "##")
(sdegeo:define-2d-contact    (find-edge-id (position (+ 50 Lg 35) tac 0)) "D")
;----- Substrate -----
(sdegeo:define-contact-set "substrate" 4.0    (color:rgb 1.0 0.0 0.0 ) "##")
(sdegeo:define-2d-contact    (find-edge-id (position (+ x2 (/ Lg 2)) (- Body) 0))
"substrate")


;----- (4). Doping -----;
;--- Channel ---
(sdedr:define-constant-profile "dopedC" "BoronActiveConcentration" C_Doping )
(sdedr:define-constant-profile-region    "RegionC" "dopedC" "Channel" )
```

```
;--- Source ---
(sdedr:define-constant-profile "dopedS" "PhosphorusActiveConcentration"
S_Doping )
(sdedr:define-constant-profile-region    "RegionS" "dopedS" "Source" )
(sdedr:define-constant-profile "dopedSC" "PhosphorusActiveConcentration"
S_Doping )
(sdedr:define-constant-profile-region    "RegionSC" "dopedSC" "SourceC" )
;--- Drain ---
(sdedr:define-constant-profile"dopedD" "PhosphorusActiveConcentration"
D_Doping )
(sdedr:define-constant-profile-region    "RegionD" "dopedD" "Drain" )
(sdedr:define-constant-profile "dopedDC" "PhosphorusActiveConcentration"
D_Doping )
(sdedr:define-constant-profile-region    "RegionDC" "dopedDC" "DrainC" )
;--- Body ---
(sdedr:define-constant-profile "dopedB" "BoronActiveConcentration" B_Doping )
(sdedr:define-constant-profile-region    "RegionB" "dopedB" "Body" )
;----- (5). Mesh -----;
;--- AllMesh ---
(sdedr:define-refinement-size "Cha_Mesh" 20 20 0 10 10 0)
(sdedr:define-refinement-material "channel_RF" "Cha_Mesh" "Silicon" )
;--- ChannelMesh ---
(sdedr:define-refinement-window "multiboxChannel" "Rectangle"
(position 25 (- 50) 0)
(position (+ 50 Lg 25) (+ tac 50) 0))
(sdedr:define-multibox-size "multiboxSizeChannel"    5 5 0 1 1 0)
(sdedr:define-multibox-placement "multiboxPlacementChannel"
"multiboxSizeChannel" "multiboxChannel")
(sdedr:define-refinement-function "multiboxPlacementChannel"
"DopingConcentration" "MaxTransDiff" 1)


;----- (6). Save (BND and CMD and rescale to nm) -----;
(sde:assign-material-and-region-names (get-body-list) )
(sdeio:save-tdr-bnd (get-body-list) "n@node@_nm.tdr")
(sdedr:write-scaled-cmd-file "n@node@_msh.cmd" nm)
```

```
(define sde:scale-tdr-bnd
(lambda (tdrin sf tdrout)
    (sde:clear)
    (sdegeo:set-default-boolean "XX")
    (sdeio:read-tdr-bnd tdrin)
    (entity:scale (get-body-list) sf)
    (sdeio:save-tdr-bnd (get-body-list) tdrout)
    )
 )
(sde:scale-tdr-bnd "n@node@_nm.tdr" nm "n@node@_bnd.tdr")
;----------------------------------------- END -------------------------------------;
```

Now, we explain the code file of **devise_dvs.cmd** item by item.

In Devise commands, the codes behind ";" and "#" are the two prompt characters as the notes by program designer which will not be executed by computer.

Silicon is selected as the active device material. The length of source and drain is set at 50 nm. To be able to create the contact boundary within 50 nm, source and drain are each composed of two extra 2D rectangles, and the length of each rectangle is 25 nm. **@Lg@ is the symbol of @variable@**, and the value can be set as the variable in **Sentaurus WorkBench (SWB)** in Fig. 2.5. The benefit in doing this is that multiple variables L_g can be assigned all at once to analyze the difference of 2D MOSFET under different L_g as shown in **SWB** of Fig. 2.7.

The thickness of gate oxide @Tox@ is also the symbol of @variable@, but in this Example 2.1, we just set a single value of 5 nm. Users can change the tox value in Workbench.

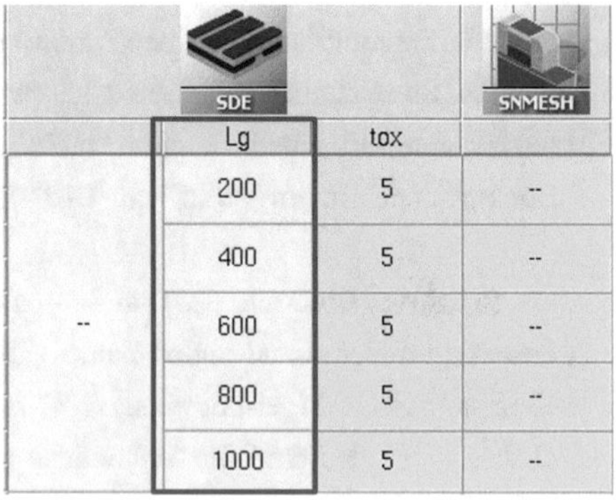

Fig. 2.7 Assigning values of @Lg@ in workbench

	Lg	tox	
	200	5	--
	400	5	--
--	600	5	--
	800	5	--
	1000	5	--

In this example, the thicknesses of gate length (L_g) and gate oxide (tox) layer are set as variables @Lg@ and @tox@, each of which is composed by two rectangles. The body is composed of one rectangle, so it is composed of a total of "**seven 2D rectangles.**"

In the same line of code, materials, number or parameters with different definitions must be separated by blanks. The universal code for parameter definition is (**define A B**), in which A represents the name declared to the computer, and B represents the value of A. The default unit of this simulation software is μm, but the unit of nm is more suitable for current semiconductor device design. Therefore, the unit of nm must be set in the parameter setting with the value of μm multiplied by 10^{-3}, defined as "(**define nm 1e-3**)." In addition, in this book, "**tac**" is defined as the thickness of active layer, "**tox**" is defined as the thickness of gate oxide, "**Body**" is defined as the thickness of body beneath active layer, "**LSDC**" is defined as the length of source/drain contact, "**LSD**" is defined as the length of source/drain, "**C_Doping**" is defined as the doping concentration of channel, "**D_Doping**" is defined as the doping concentration of drain, "**S_Doping**" is defined as the doping concentration of source, and "**B_Doping**" is defined as the doping concentration of body.

And then the algebra of x1, x2, x3… is used for representing all values to easily facilitate the code adjustment during lots of parameter modification in the future and the device structure scaling. It is revealed in Fig. 2.6 that the algebra of every coordinate (x1, x2, …) is the sum of the previous algebra and the previously defined parameter. For example, x2 is equaled to x1 plus LSDC.

It is noteworthy that, in this Sentaurus TCAD software, the form of universal codes of addition/subtraction/multiplication/division is (operator **A B**). Its mathematical meaning is the operator calculation of A with respect to B. For example, (+ 10 5) means $10 + 5$, (− 10 5) means $10 - 5$, and (− 10 5 2) means $10 - 5 - 2$. Attentions must be paid to the sequence of multiplication and division operation. For example, (/10 5 2) means (10/5)/2, and it is because the order of calculation in this program is from left to right.

If there is another parenthesis behind 10, the calculation order can be changed. For example, (/10 (/5 2)) means 10/(5/2). And then, we enter the section of device structure establishment.

In this example, the device structure is established via rectangle stacking. It is shown in Fig. 2.6 that a total of seven rectangles are used in this example. The benefit of this method is that it can easily establish the desired structure. However, there will be the problem of computer failed to identify the intersection or overlapped region of rectangles. "**ABA**" is exactly the code to deal with this issue. It means that the new rectangle will replace the old rectangle at the region with overlapping new and old rectangles, and the material property of this region will be determined by the new rectangle as shown in Fig. 2.8.

As for the device structure establishment, the establishment of rectangles will be declared by sdegeo. The rectangle refers to the 2D rectangular rectangle, and the size of rectangle is determined by the diagonal.

As shown in Fig. 2.9, the size of rectangle can be determined by the assigned coordinates of A and B. For example, (**sdegeo:create-rectangle (position 0 0 0)**

Fig. 2.8 Illustration of "ABA" command. It is very useful in 3D FET; **a** is suitable for gate-all-around FET, and **b** is suitable for FinFET

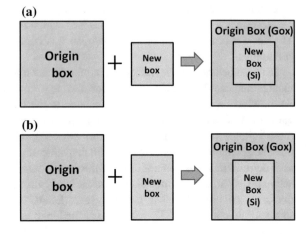

Fig. 2.9 Illustration of rectangle establishment, vector from point A to point B

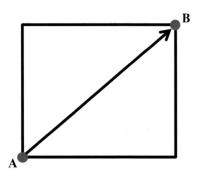

(position x1 y2 0) "Silicon" "SourceC"), where the coordinate of A is (0, 0, 0), and the coordinate of B is (x1, y2, 0). After the coordinates of A and B are defined, the material of this rectangle will be declared, and in this example, the material is silicon. After material definition, the name of this rectangle will be declared, and it is defined as SourceC in this example, which represents Source Contact.

Then the six rectangles of **SourceC, Source, Channel, Drain, DrainC, and gate oxide** will be properly established from left to right in accordance with Fig. 2.6. After the rectangles required by **active layer** are established, the next thing is to establish the rectangles for body and gate oxide. In the end, the voltage must be applied. When a voltage is applied to gate G, drain D, source S, and body B, the entire rectangle of electrode can be regarded as a complete conductor, and all electrodes are at the same potential. The definition of gate electrode on the gate oxide will form an equipotential surface without the need for additionally defined gate material. The only thing to be noted is that the **work function of metal gate** must be properly defined in order to achieve the expected threshold voltage (V_{th}). And then, all contacts required by the device must be defined.

Contact code: **define-contact-set "G"** is for assigning the name of contact which can be self-defined. **(color:rgb 1.0 0.0 0.0) "##"** is for defining the color and form of contact, and here it can be set as default. This example is a 2D simulation, so the 2D contact needs to be established such as define-2d-contact. It is noteworthy that, as described in previous section, only the electrode equipotential lines need to be defined (equipotential surface in 3D structure), and **there is unnecessary for defining physical material and physical space of contact**. By assigning a point to the defined equipotential line, the program will automatically stretch to the left and to the right from that point until reaching the boundaries of this 2D rectangle, and this extended line is the closed electrode equipotential line. The code is **find-edge-id (position (+ x2 (/ Lg 2)) y3 0)) "G"**. It is noteworthy that the name of electrode must be given once here, and the definition G means "gate contact electrode." And then, the contact electrodes of source, drain, and substrate must be established in proper order.

After all required contacts are established, the next thing is to define the doping styles and doping concentrations of seven rectangles.

The rule of doping establishment is to determine the style and concentration of doping before putting into the previously established rectangles to complete the doping process. The code of **"define-constant-profile"** refers to a doping with fixed style and concentration which does not take into consideration the concentration gradient.

"dopedC" refers to the names of this doping style and concentration, which can be assigned based on personal preference. **"BoronActiveConcentration"** refers to a dopant which is already activated, and here it is boron. The **C_Doping** refers to the concentration of this doping, and the value of C_Doping has been assigned during parameter definition.

Code: **define-constant-profile-region** refers to putting the pre-defined doping style and concentration into a fixed region without considering the situation of dopant diffusion. "RegionC" refers to the name of this action which can be set in accordance with personal reference. The content of this action is to combine all codes showing up afterward, so if **"RegionC" is followed by "dopedC" and "Channel," it refers to putting dopedC into the rectangle of Channel, and this action is called RegionC**. This is how the doping of seven rectangles is composed.

After the definition of doping is completed, the mesh of mathematical calculation should be determined. The simulation calculation of semiconductor device must be analyzed by various physics formula, the most fundamental of which are the Poisson equation to determine electrical potential and the continuity equation to determine carrier concentration. The electrical properties of semiconductor device must be based on the simultaneous solution, such as Poisson equation, continuity equation, and transport equation. The location of such solution is the intersection of mesh as shown in Fig. 2.10. However, there cannot be infinite number of solutions for semiconductor device, so Newton interpolation method is used for approximation between points. However, the approximation by interpolation is not a proper

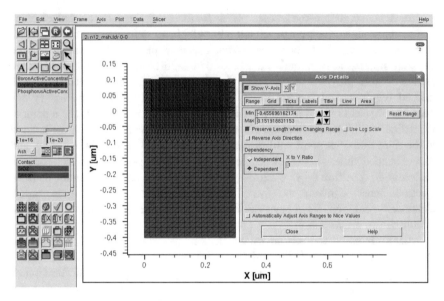

Fig. 2.10 Actual example of mesh. The active layer is designed with denser mesh, and the substrate layer is designed with less dense mesh in order to obtain optimized simulation design efficiency

option for regions with large concentration gradient or electric field variation. These regions must be analyzed by more precise solutions, so the meshes in these regions must be denser.

The logic behind the establishment of mesh is: First a comprehensive mesh is assigned. Since the locations of solutions are the intersections of mesh, a dense mesh will be assigned to the region with large concentration gradient, large electric field variation, or significant impact on electrical properties (such as the active region). Therefore, an **AllMesh** is assigned first. **define-refinement-size** refers to the definition of distance between points. With varying geometric shapes of semiconductor device, there will be different distributions of dopant concentrations such that the program will automatically adjust the distance on the boundary as long as the maximum and minimum values are declared to the computer. **"Cha_Mesh"** refers to the name of aforementioned action, which can be determined based on personal preference. Among the next six numbers, the first three of them refer to the maximum values along the three directions of X-axis, Y-axis, and Z-axis, and the next three of them refer to the minimum values along the three directions of X-axis, Y-axis, and Z-axis. In this case, these numbers are **(20 20 0 10 10 0)**, because the maximum value of X is 20 nm and the minimum value of X is 10 nm, and the same shall apply to Y, so the maximum and minimum values of Z are set to be zero, because this example is 2D device.

The mesh established earlier can be placed in specific region or specific material, and here it is placed in the specific material by the code: **define-refinement-material**. **"channel_RF"** is the name of this action, which can be determined based on personal preference. However, it is better to be in accordance with the syntax suggested in this book to avoid unnecessary error. This action is to combine the following code, so if "channel_RF" is followed by "Cha_Mesh" and "Silicon," it refers to the action of placing Cha_Mesh into the material of silicon. This action is called channel_RF (i.e., channel refinement).

After the comprehensive mesh is established, a denser mesh will be established in the active layer. The method for doing this is to establish a mesh rectangle and then put it in the designated region. The code: **define-refinement-window** is for establishing the mesh rectangle. The name of this action is **multiboxChannel**, which can be determined based on personal preference, yet it is better to be in accordance with the syntax suggested in this book to avoid unnecessary error. "Rectangle" refers to the establishment of a 2D mesh rectangle based on the method of determining the size of rectangle by diagonal before assigning the maximum and minimum values along the three directions of X-axis, Y-axis, and Z-axis. And the code: **multiboxPlacementChannel** is the name of this action which can be determined based on personal preference. This action is to integrate the following code, so if **"multiboxPlacementChannel" is followed by "multiboxSizeChannel" and "multiboxChannel," it refers to placing multiboxSizeChannel into multiboxChannel, and this action is called multiboxPlacementChannel**.

The denser mesh will be stretched across the region with larger variation of concentration gradient, so here another line of code will be added for the denser mesh to be stretched toward the depletion region. The code is define-refinement-function. This action will take place in multiboxPlacementChannel with extension in accordance with DopingConcentration. The MaxTransDiff refers to the degree of extension, which is one. There has more selection of MaxTransDiff degree; readers can refer the TCAD manual.

And next the file is saved with the code as shown in the last part of code:

The part can be used as default. Among them, assign-material-and-region-names is for saving the previously established materials and names, and the line of write-scaled-cmd-file is for saving the aforementioned scale. With the default unit of program being μm, X-axis, Y-axis, and Z-axis must be multiplied by 10^{-3} and the unit should be converted to nm. The following scale-tdr-bnd saves all files as boundary format. It will be used by the following tool. All codes required by SDE are hereby completed.

The second tool in the tool column is **"SNMESH"** tool in SWB, which is mainly used for establishing the mesh required by device simulation, and the mesh code has been written in the commands of SDE. So we only have to set **SNMESH to access the commands of SDE** as shown in Fig. 2.11.

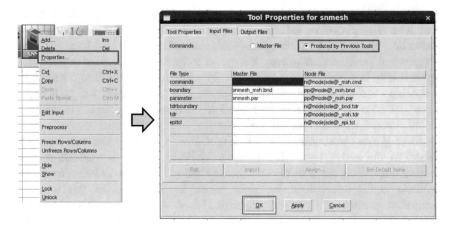

Fig. 2.11 Modification of properties of SNMESH

By right-clicking on Properties on **SNMESH tool** to enter the input files and selecting Produced by Previous Tools, **SNMESH** will automatically grab the codes of previous **SDE**, and there is no need for entering any code into **SNMESH**.

By now a complete 2D MOSFET device structure has been established, yet it has not been given any physics model. Therefore, the next tool: **SDEVICE** is used for applying the physics model and mathematic model of this semiconductor device and the conditions of threshold voltage (V).

2. **SDEVICE → dessis_des.cmd**

SDEVICE tool commands (code file is **dessis_des.cmd**) can be divided into six parts.

(1) **Electrode**
(2) **File**
(3) **Physics**
(4) **Math**
(5) **Plot**
(6) **Solve**

The complete code of **SDEVICE** is as shown below. It is noteworthy that in the commands of **SDEVICE, the symbol * or # indicates that the following line of code is the prompt character for program designers to keep notes, and it will not be executed by the computer**.

The **SDEVICE** codes of this 2D MOSFET example are as shown below:

```
#------------    dessis_des.cmd    ---------------#
Electrode{
{name="D" voltage=0.0}
{name="S" voltage=0.0}
{name="G" voltage=0.0 WorkFunction=@WK@}
}
File{
        Grid="@tdr@"
        Plot="@tdrdat@"
        Current="@plot@"
        Output="@log@"
        parameter="@parameter@"
}
Physics{
    Mobility( DopingDep HighFieldSaturation Enormal )
    EffectiveIntrinsicDensity( OldSlotboom )
    Recombination( SRH(DopingDep) )
    eQuantumPotential
}
Math{
    -CheckUndefinedModels
    Number_Of_Threads=4
    Extrapolate
    Derivatives
    * Avalderivatives
    RelErrControl
    Digits=5
    ErRef(electron)=1.e10
    ErRef(hole)=1.e10
    Notdamped=50
    Iterations=20
    Directcurrent
    Method=ParDiSo
    Parallel= 2
    -VoronoiFaceBoxMethod
```

```
        NaturalBoxMethod
    }
    Plot{
        eDensity hDensity
        eCurrent hCurrent
        TotalCurrent/Vector eCurrent/Vector hCurrent/Vector
        eMobility hMobility
        eVelocity hVelocity
        eEnormal hEnormal
        ElectricField/Vector Potential SpaceCharge
        eQuasiFermi hQuasiFermi
        Potential Doping SpaceCharge
        SRH Auger
        AvalancheGeneration
        DonorConcentration AcceptorConcentration
        Doping
          eGradQuasiFermi/Vector hGradQuasiFermi/Vector
          eEparallel hEparalllel
          BandGap
          BandGapNarrowing
          Affinity
          ConductionBand ValenceBand
          eQuantumPotential
    }
    Solve {
        Coupled ( Iterations= 150){ Poisson eQuantumPotential }
        Coupled { Poisson eQuantumPotential Electron Hole }
        Quasistationary(
            InitialStep= 1e-3 Increment= 1.2
            MinStep= 1e-12 MaxStep= 0.95
            Goal { Name= "D" Voltage=@Vd@ }
        ){ Coupled { Poisson eQuantumPotential Electron Hole } }
            Quasistationary(
            InitialStep= 1e-3 Increment= 1.2
            MinStep= 1e-12 MaxStep= 0.02
```

```
      Goal { Name= "G" Voltage=@Vg@ }
      DoZero
     ){ Coupled { Poisson eQuantumPotential Electron Hole } }
   }
#---------------------------------   END   -------------------------------------#
```

At first, all the conditions of voltage (V_d, V_s, and V_g) will be defined together with the gate. The metal gate work function (WorkFunction(V)) sets as a variable @WK@.

The name of each electrode terminal must be identical to the name in SDE tool, or the program cannot be executed due to interpretation failure. The work function will be set as variables to be defined in workbench to facilitate the calibration of threshold voltage (V_{th}).

The next step is to set up the files to be read from the previous tool during the operation of **SDEVICE**. This part is based on default value such that it cannot be modified without permission, because the file name must be in compliance with the program regulation.

And then, we need to tell the computer what are the physics formula to be substituted into this simulation calculation, such as the recombination model or some quantum modification models.

Among them, the **eQuantumPotential** is the quantum modification item with respect to the density of state of electron. When the device dimension is very small, some quantum modification items must be added for the simulation results to be closer to the real condition. The default setting of this program is based on the most complete physics model, so please read the manual thoroughly before making any modification.

After assigning physics model, the next step is the assignment of mathematical model, which can be based on default setting. The default setting of this program is based on the most complete mathematic model, so please read the manual thoroughly before making any modification.

It is noteworthy that **Iterations = 20**, it indicates the number of points to be given during approximation by Newton interpolation method between two mesh points, and 20 points are given here. Usually the less dense mesh will be set up first before the mesh of important simulation step and crucial region (such as active layer) can be optimized during simulation process in order to be in compliance with the correct electrical and physical properties of the device. Therefore, this part of setting can be determined based on personal preference. Insufficient number of points given here will lead to divergence of simulation calculation, yet excessive number of points given will lead to prolonged simulation calculation.

The next step is to tell the computer what are the diagrams to be extracted, such as energy band diagram and carrier distribution diagram, which can be deleted based on personal preference.

In the end, **SDEVICE** tool should tell the computer what are the calculations to be done by the combination of aforementioned physics model and mathematic model (such as Poisson equation).

It is revealed in the code that electron and hole must be substituted in Poisson equation for solution, and eQuantumPotential must be added for correction. The reader should find out that in this book, the drain (D) voltage is addressed before gate (G) voltage (V_g). It should be noted that the voltage mentioned earlier is regarded as a constant value, and the voltage mentioned later is for sweeping action based on the range set in accordance with workbench. Another point to be noted is that InitialStep = $1e^{-3}$ refers to the first point to be calculated will be $1e^{-3}$ unit away from threshold voltage. If it cannot be converged, the voltage pitch will be reduced to continue with the search for value that can be converged. The minimum value to be reduced is $1e^{-12}$ unit, and the maximum value is 0.02 unit, which can be modified based on personal preference.

3. INSPECT → inspect_ins.cmd

In the end, it comes to the part of **INSPECT** tool, which should be set up in accordance with the default value without the need for any modification. This is because INSPECT will only affect what are the parameters to be extracted. The author has suggested that this function should be for reference only. If there is any need for cautious data analysis, the data should be extracted and analyzed by professional engineering and scientific application graphic software such as SigmaPlot and Origin. The **symbol #** indicates that the following line of code is the prompt character for program designers to keep notes, and it will not be executed by the computer.

The command lines of INSPECT are usually using default file as shown below:

```
#-------    inspect_ins.cmd ------------------------#
#                Script file designed to compute     : #
#                   * The threshold voltage          : VT #
#                   * The transconductance        : gm#
#----------------------------------------------------------#
if { ! [catch {open n@previous@_ins.log w} log_file] } {
      set fileId stdout
}
puts $log_file " "
puts $log_file "               ----------------------------------- "
puts $log_file "               Values of the extracted Parameters : "
puts $log_file "               ----------------------------------- "
puts $log_file " "
puts $log_file " "
set   DATE     [ exec   date ]
set   WORK     [ exec pwd    ]
puts $log_file "    Date         : $DATE "
puts $log_file "    Directory : $WORK "
puts $log_file " "
puts $log_file " "
#                          idvgs=y(x) ;    vgsvgs=x(x) ;                        #
set out_file n@previous@_des
proj_load "${out_file}.plt"
# ---------------------------------------------------------------------- #
# I)   VT = Xintercept(maxslope(ID[VGS]))   or   VT = VGS( IDS= 0.1 ua/um ) #
# ---------------------------------------------------------------------- #
cv_create     idvgs    "${out_file} G OuterVoltage" "${out_file} D TotalCurrent"
cv_create     vdsvgs   "${out_file} G OuterVoltage" "${out_file} D OuterVoltage"
#.................................................. #
# 1) VT extracted as the intersection point with the X axis at the point #
#      where the id(vgs) slope reaches its maximum : #
#.................................................. #
set VT1     [ f_VT1 idvgs ]
#...........................................              #
# 2) Printing of the whole set of extracted values (std output) :   #
```

```
#...............................................                 #
puts $log_file "Threshold      voltage VT1     = $VT1   Volts"
puts $log_file " "
#...................................................       #
# 3) Initialization and display of curves on the main Inspect screen   :   #
# ...................................................       #
cv_display       idvgs
cv_lineStyle    idvgs    solid
cv_lineColor    idvgs    red
# ------------------------------------------------------------------------ #
# II)                    gm =   maxslope((ID[VGS])              #
# ------------------------------------------------------------------------ #
set gm       [ f_gm idvgs ]
puts $log_file " "
puts $log_file "Transconductance gm               = $gm   A/V"
puts $log_file " "
set ioff   [ cv_compute "vecmin(<idvgs>)" A A A A ]
puts $log_file " "
puts $log_file "Current ioff              = $ioff   A"
puts $log_file " "
set isat   [ cv_compute "vecmax(<idvgs>)" A A A A ]
puts $log_file " "
puts $log_file "Current isat              = $isat   A"
puts $log_file " "
set rout   [ cv_compute "Rout(<idvgs>)" A A A A ]
puts $log_file " "
puts $log_file "Resistant rout             = $rout   A"
puts $log_file " "
cv_createWithFormula logcurve "log10(<idvgs>)" A A A A
cv_createWithFormula difflog "diff(<logcurve>)" A A A A
set sslop [ cv_compute "1/vecmax(<difflog>)" A A A A ]
puts $log_file " "
puts $log_file "sub solp            = $sslop   A/V"
puts $log_file " "
### Puting into Family Table #####
```

```
ft_scalar VT $VT1
ft_scalar gmax $gm
ft_scalar ioff $ioff
ft_scalar isat $isat
ft_scalar sslop $sslop
ft_scalar rout $rout
close $log_file
#------------------------------- END ------------------------------------#
```

As for the extraction of V_{th}, the drain current is to be collected with this example based on nFET, and it is set as D. If the simulation is for pFET, it should be changed to S. As for the extraction of SS, if it is for pFET, diff(<logcurve>) must be multiplied by (−1) before the current of pFET is in opposite direction to the current of nFET.

By now all codes have been entered, and in the end, all parameters of workbench should be set, and please remember to enter all variables to be included in the workbench as shown in Fig. 2.12. The setting method is as shown in Fig. 2.13.

First, right-click to select **Add** in the lower column of Tool, and then enter the name of variable in Parameter and the value of variable in Default Value. The values can be deleted or changed by right-clicking **Edit Values** in the lower column of Tool as shown in Fig. 2.14.

After the variables are completed, the project name can be selected, and **run** can be clicked to start the simulation calculation as shown in Fig. 2.15.

It is shown in Fig. 2.16 that the name of project will turn **yellow** after the completion of simulation, indicating a successful simulation calculation; if it turns **red**, it means the simulation cannot be converged or has some syntax error. User can check error message in log file. Also the mesh calculation may need to be adjusted for convergence.

SDE			SNMESH		SDEVICE			INSPECT
Lg	tox				WK	Vg	Vd	
200	5	--	--		4.8	3	0.1 3	-- --
400	5	--	--		4.8	3	0.1 3	-- --
600	5	--	--		4.8	3	0.1 3	-- --
800	5	--	--		4.8	3	0.1 3	-- --
1000	5	--	--		4.8	3	0.1 3	-- --

Fig. 2.12 Important variables being set up of each tool in the SWB

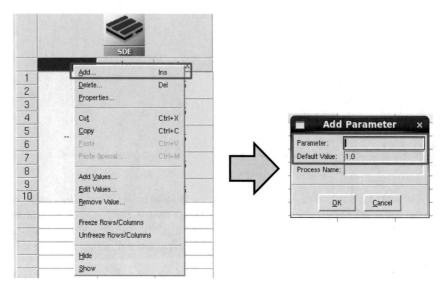

Fig. 2.13 Method to set up the variables in the workbench

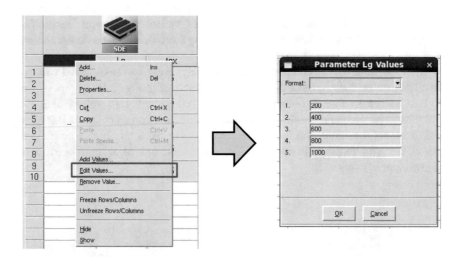

Fig. 2.14 Method for modifying variables of workbench

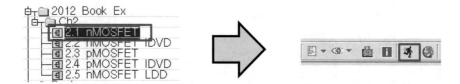

Fig. 2.15 Starting the simulation calculation, pressing **RUN icon**

Fig. 2.16 Icons of simulation successful in *yellow* or failed in *red*

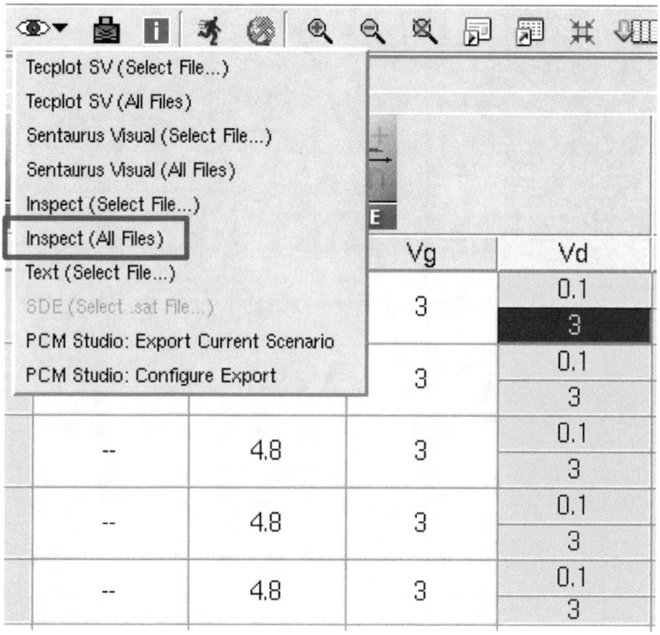

		Vg	Vd
		3	0.1
			3
		3	0.1
			3
--	4.8	3	0.1
			3
--	4.8	3	0.1
			3
--	4.8	3	0.1
			3

Tecplot SV (Select File...)
Tecplot SV (All Files)
Sentaurus Visual (Select File...)
Sentaurus Visual (All Files)
Inspect (Select File...)
Inspect (All Files)
Text (Select File...)
SDE (Select .sat File...)
PCM Studio: Export Current Scenario
PCM Studio: Configure Export

Fig. 2.17 Open Inspect tool to show electrical property

After the completion of simulation, the next step uses **INSPECT tool to analyze electrical properties** as shown in Fig. 2.17, where the node of the electrical property should be right-clicked and the eye icon on the visualize bar should be clicked to select Inspect (All Files).

The selected interface is as shown in Fig. 2.18. The node to be inspected should be selected on the datasheet of workbench, and the set electrode and the electrical property to be inspected should be selected before being put on the preset axis. For operation example: **G → OuterVoltage → To X-Axis; D → TotalCurrent → To Left Y-Axis**. In the end, the Y-axis on the topmost tool bar should be changed to be displayed in log scale to lead us to the frequently seen I_d–V_g curve.

In addition, the data of I_d–V_g curve can be exported in txt file format to be analyzed by other professional engineering and scientific application graphics software such as **SigmaPlot** and **Origin**. As indicated in Fig. 2.19, the data of

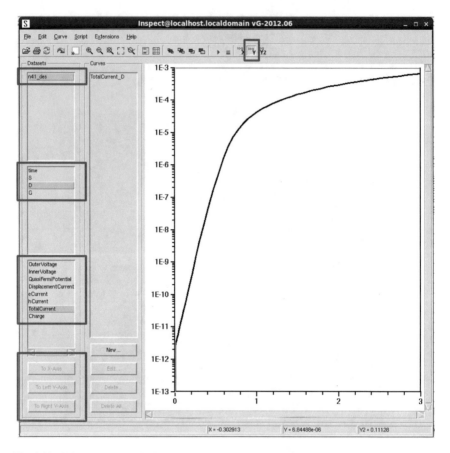

Fig. 2.18 Using Inspect tool plots electrical properties display and analysis. The plot shows a typical I_d–V_g transfer curve of nMOSFET

electrical properties can be exported by clicking the **File $\rightarrow$ Export $\rightarrow$ csv or txt format** on the upper left corner. The data of electrical properties is exported in the format of csv or txt file as shown on the right side of Fig. 2.19, where the data of x is V_g, and the data of y is I_d.

Except for the electrical property diagram, other important physical properties, such as electric field, electrostatic potential, and charge concentration, need to be examined during the analysis of semiconductor device. As shown in Fig. 2.20, the node of electrical property to be examined should be right-clicked, and the eye icon on the Visualize bar should be selected to access the drop-down menu of visualization software. And then, we should select Sentaurus Visual (All Files).

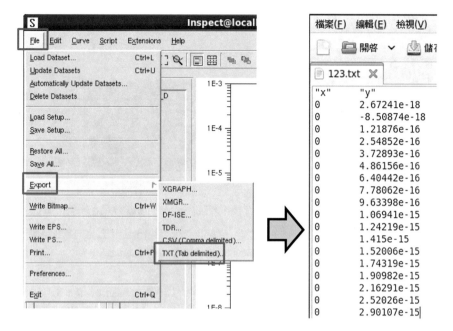

Fig. 2.19 Exported data of electrical properties is in the txt file format (ex: IdVg.txt)

Fig. 2.20 Select Sentaurus
Visual to examine physical
properties

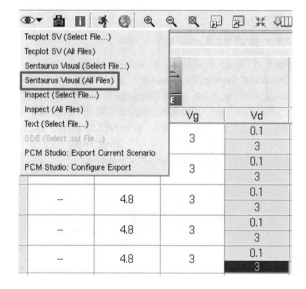

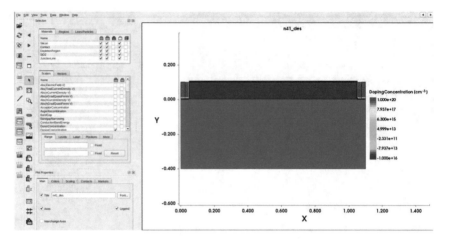

Fig. 2.21 Sentaurus Visual interface (window-based user-friendly interface)

Figure 2.21 is the Sentaurus Visual interface. The Selection on the left is for selecting the material to be displayed and its mesh. In addition, the physical property to be inspected can be selected on the lower part of screen such as: energy band diagram and carrier distribution of electrons and holes. If either X-axis or Y-axis is to be fixed to observe the variation of physical property along with the other axis, the icons in the red frame on the right can be selected. For example, fixing X-axis at the position of 0.5 will allow us to observe the variation of physical property along Y-axis with X = 0.5.

Special Note: FAQ and Troubleshooting
The most frequently seen problem is that the value does not converge during the simulation, and the error message is as shown in Fig. 2.22.

Most of these problems are due to the difficulty in mesh calculation which has resulted in singular points generation at the location with large concentration variation gradient thus causing diversion. The better mesh code which is less vulnerable to diversion as shown below:

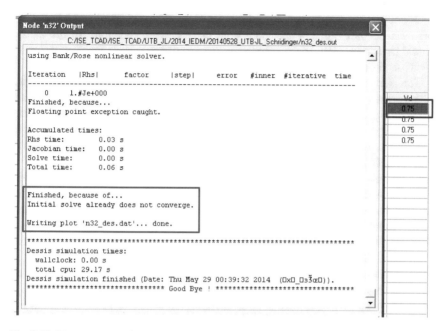

Fig. 2.22 Error message of the value of simulation does not converge

;------   AllMesh   -----;

(isedr:define-refinement-size "Cha_Mesh" 5 5 5 1 1 1)

(isedr:define-refinement-material "channel_RF" "Cha_Mesh" "Silicon")

The Channel width here is 5 nm, so the initial rough cutting range is 5 nm → 1 nm for the cutting of the entire silicon.

;------ Channel Mesh example of 3D FinFET -----;

;--- AllMesh ---;

(sdedr:define-refinement-size "Cha_Mesh" 5 5 5 1 1 1)

(sdedr:define-refinement-material "channel_RF" "Cha_Mesh" "Silicon")

;--- ChannelMesh ---;

(sdedr:define-refinement-window "multiboxChannel" "Cuboid"

(position x1 0 0)   (position x4 y1 z1))

(sdedr:define-multibox-size "multiboxSizeChannel"   2 2 2 2 2 2)

(sdedr:define-multibox-placement "multiboxPlacementChannel"

"multiboxSizeChannel" "multiboxChannel")

(sdedr:define-refinement-function "multiboxPlacementChannel"

"DopingConcentration" "MaxTransDiff" 1)

The following fine cutting is mainly placed inside the channel because there is large carrier variation gradient and large electric field variation. So the cutting should be 1 nm → 0.5 nm or 1 nm → 2 nm. In this book, it is suggested that mesh variation should be kept within 100%.

The max value and min value of mesh should be determined in coordination with device dimension from high to low. The divergence will most likely to take place in the region with large concentration gradient and electric field variation. This is because the Newton interpolation method is used for approximation at the intersection of mesh, and the value is confirmed by left limit and right limit approach. If the values of left limit and right limit do not match, the simulation result will diverge. Therefore, smaller mesh should be assigned to the location with large concentration gradient or large electric field variation. The device for the first run can be assigned with a larger mesh or fewer elements to see if the electric properties are as expected and finer segmentation can be applied for observation of electric field distribution or carrier distribution. It still cannot converge after mesh adjustment, fine-tuning of **workfunction** can be considered. For example, if the original value is **4.6 eV, we can try with 4.601 eV. An additional 0.001 eV will not lead to too much impact on V_{th}, but it can help with convergence**. Voltage fine-tuning can also be applied such as changing $V_D = 1$ V to $V_D = 1.01$ V. In this book, it is suggested that the adjustment shall not exceed **2%** (Fig. 2.23).

The analysis of physical properties of 2D nMOSFET based on **Sentaurus Visual interface** is as shown in Fig. 2.24.

2D-nMOSFET

Fig. 2.23 I_d–V_g curves of simulation of 2D nMOSFET. Some descriptions are added by PowerPoint after snapshot from Inspect. It reveals that the 2D nMOSFET with $L_g < 600$ nm suffers severe short-channel effect (SCE)

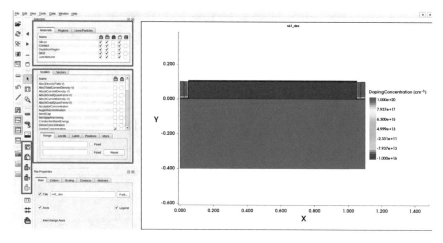

Fig. 2.24 Use Sentaurus Visual interface. Important device physical properties can be visualized and analyzed

Figure 2.24 is the Sentaurus Visual interface. The Selection on the left is for selecting the material to be displayed and its mesh. In addition, the physical property to be inspected can be selected on the lower part of screen such as: energy band diagram and carrier distribution of electrons and holes.

The electron concentration distribution, electric field distribution, electro-static potential distribution, and energy band diagram along the channel direction are as shown in Figs. 2.25, 2.26, 2.27, and 2.28 based on the conditions of L_g = 1000 nm, V_d = 3 V, and V_g = 3 V.

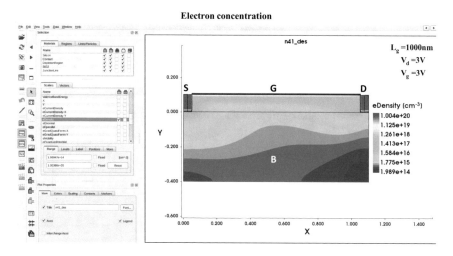

Fig. 2.25 Electron concentration distribution of the simulation of 2D n-type semiconductor device

Electric Field

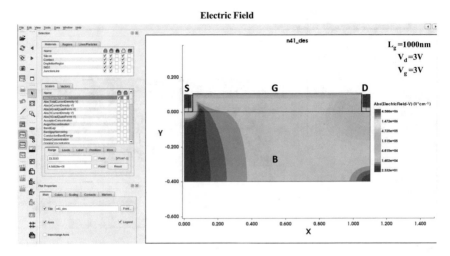

Fig. 2.26 Electric field distribution of the simulation of 2D n-type semiconductor device

Electric Field

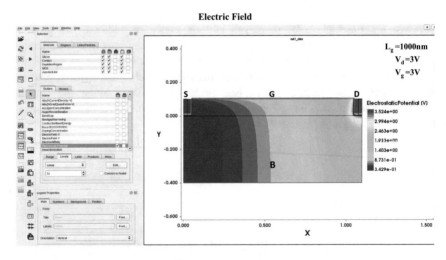

Fig. 2.27 Electrostatic potential distribution of the simulation of 2D n-type semiconductor device

Band Diagram

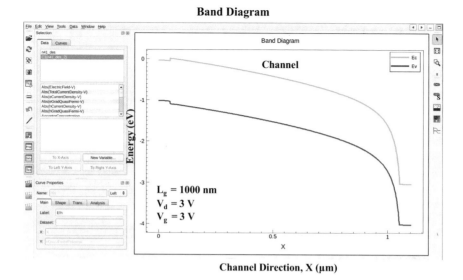

Channel Direction, X (μm)

Fig. 2.28 Energy band diagram along the channel direction of the simulation of 2D n-type semiconductor device

2.3   [Example 2.2] 2D n-Type MOSFET with I_d–V_d Characteristics Simulation

This I_d–V_d example is very similar to Example 2.1, only has difference in electrodes and its bias setting.

The following three main program code files are all based on Synopsys Sentaurus TCAD 2014 version.

1. SDE—devise_dvs.cmd

This is the best example of 2D nMOSFET I_d–V_d.

The line of code following; **is the prompt character for program designer to take note such that it will not be executed by the computer.**

```
;----- parameter -----;
(define Lg @Lg@)
(define tox @tox@)
(define tac 100)
(define Body 400)
(define LSDC 25)
(define LSD 25)
(define C_Doping 1e16)
(define D_Doping 1e20)
(define S_Doping 1e20)
(define B_Doping 1e15)
(define nm 1e-3)
(define x1 LSDC)
(define x2 (+ x1 LSD))
(define x3 (+ x2 Lg))
(define x4 (+ x3 LSD))
(define x5 (+ x4 LSDC))
(define y1 (- Body))
(define y2 tac)
(define y3 (+ tac tox))
;----- Structure -----;
"ABA"
;--- source ---
(sdegeo:create-rectangle
   (position      0 0 0)
   (position      x1 y2 0)   "Silicon" "SourceC" )
(sdegeo:create-rectangle
   (position      x1 0 0)
   (position      x2 y2 0)   "Silicon" "Source" )
;--- Channel ---
(sdegeo:create-rectangle
   (position      x2 0 0)
   (position      x3 y2 0)   "Silicon" "Channel" )
;--- Drain ---
(sdegeo:create-rectangle
```

```
    (position      x3 0 0)
    (position      x4 y2 0)    "Silicon" "Drain" )
(sdegeo:create-rectangle
    (position      x4 0 0)
    (position      x5 y2 0)    "Silicon" "DrainC" )
;--- Body ---
(sdegeo:create-rectangle
    (position      0 0 0)
    (position      x5 y1 0)    "Silicon" "Body" )
;--- Gate oxide ---
(sdegeo:create-rectangle
    (position      x2 y2 0)
    (position      x3 y3 0)    "SiO2" "Gateoxide" )
;----------------------- Contact ----------------------------;
;----- Gate -----
(sdegeo:define-contact-set "G"
    4.0    (color:rgb 1.0 0.0 0.0 ) "##")
(sdegeo:define-2d-contact    (find-edge-id (position (+ x2 (/ Lg 2)) y3 0)) "G")
;----- Source -----
(sdegeo:define-contact-set "S"
    4.0    (color:rgb 1.0 0.0 0.0 ) "##")
(sdegeo:define-2d-contact    (find-edge-id (position 10 tac 0)) "S")
;----- Drain -----
(sdegeo:define-contact-set "D"
    4.0    (color:rgb 1.0 0.0 0.0 ) "##")
(sdegeo:define-2d-contact    (find-edge-id (position (+ 50 Lg 35) tac 0)) "D")
;----- Substrate -----
(sdegeo:define-contact-set "substrate"
    4.0    (color:rgb 1.0 0.0 0.0 ) "##")
(sdegeo:define-2d-contact    (find-edge-id (position (+ x2 (/ Lg 2)) (- Body) 0))
"substrate")
;--------------- Doping ---------------;
;--- Channel ---
(sdedr:define-constant-profile "dopedC" "BoronActiveConcentration" C_Doping )
(sdedr:define-constant-profile-region    "RegionC" "dopedC" "Channel" )
```

;--- Source ---
(sdedr:define-constant-profile "dopedS" "PhosphorusActiveConcentration"
S_Doping)
(sdedr:define-constant-profile-region   "RegionS" "dopedS" "Source")
(sdedr:define-constant-profile "dopedSC" "PhosphorusActiveConcentration"
S_Doping)
(sdedr:define-constant-profile-region   "RegionSC" "dopedSC" "SourceC")
;--- Drain ---
(sdedr:define-constant-profile "dopedD" "PhosphorusActiveConcentration"
D_Doping)
(sdedr:define-constant-profile-region   "RegionD" "dopedD" "Drain")
(sdedr:define-constant-profile "dopedDC" "PhosphorusActiveConcentration"
D_Doping)
(sdedr:define-constant-profile-region   "RegionDC" "dopedDC" "DrainC")
;--- Body ---
(sdedr:define-constant-profile "dopedB" "BoronActiveConcentration" B_Doping)
(sdedr:define-constant-profile-region   "RegionB" "dopedB" "Body")
;---------------- Mesh ------------------;
;--- AllMesh ---
(sdedr:define-refinement-size "Cha_Mesh" 20 20 0 10 10 0)
(sdedr:define-refinement-material "channel_RF" "Cha_Mesh" "Silicon")
;--- ChannelMesh ---
(sdedr:define-refinement-window "multiboxChannel" "Rectangle"
(position 25 (- 50) 0)
(position (+ 50 Lg 25) (+ tac 50) 0))
(sdedr:define-multibox-size "multiboxSizeChannel"   5 5 0 1 1 0)
(sdedr:define-multibox-placement "multiboxPlacementChannel"
"multiboxSizeChannel" "multiboxChannel")
(sdedr:define-refinement-function "multiboxPlacementChannel"
"DopingConcentration" "MaxTransDiff" 1)
;----------------- Save BND and CMD and rescale to nm -----------------------;
(sde:assign-material-and-region-names (get-body-list))
(sdeio:save-tdr-bnd (get-body-list) "n@node@_nm.tdr")
(sdedr:write-scaled-cmd-file "n@node@_msh.cmd" nm)
(define sde:scale-tdr-bnd

```
        (lambda (tdrin sf tdrout)
          (sde:clear)
          (sdegeo:set-default-boolean "XX")
          (sdeio:read-tdr-bnd tdrin)
          (entity:scale (get-body-list) sf)
          (sdeio:save-tdr-bnd (get-body-list) tdrout)
          )
        )
      (sde:scale-tdr-bnd "n@node@_nm.tdr" nm "n@node@_bnd.tdr")
      ;------------------ END ------------------;
```

2 SDVICE—dessis_des.cmd

The line of code following # and * are the prompt character for program designer to take note such that it will not be executed by the computer.

```
Electrode{
{name="D" voltage=0.0}
{name="S" voltage=0.0}
{name="G" voltage=0.0 WorkFunction=@WK@}
}
File{
        Grid="@tdr@"
        Plot="@tdrdat@"
        Current="@plot@"
        Output="@log@"
        parameter="@parameter@"
}
Physics{
    Mobility( DopingDep HighFieldSaturation Enormal )
    EffectiveIntrinsicDensity( OldSlotboom )
    Recombination( SRH(DopingDep) )
    eQuantumPotential
}
Math{
    -CheckUndefinedModels
    Number_Of_Threads=4
    Extrapolate
    Derivatives
    * Avalderivatives
    RelErrControl
    Digits=5
```

```
          ErRef(electron)=1.e10
          ErRef(hole)=1.e10
          Notdamped=50
          Iterations=20
          Directcurrent
          Method=ParDiSo
          Parallel= 2
     *-VoronoiFaceBoxMethod
          NaturalBoxMethod

     }
     Plot{
          eDensity hDensity
          eCurrent hCurrent
          TotalCurrent/Vector eCurrent/Vector hCurrent/Vector
          eMobility hMobility
          eVelocity hVelocity
          eEnormal hEnormal
          ElectricField/Vector Potential SpaceCharge
          eQuasiFermi hQuasiFermi
          Potential Doping SpaceCharge
          SRH Auger
          AvalancheGeneration
          DonorConcentration AcceptorConcentration
          Doping
           eGradQuasiFermi/Vector hGradQuasiFermi/Vector
           eEparallel hEparalllel
           BandGap
           BandGapNarrowing
           Affinity
           ConductionBand ValenceBand
           eQuantumPotential

     }
     Solve {
          Coupled ( Iterations= 150){ Poisson eQuantumPotential }
          Coupled { Poisson eQuantumPotential Electron Hole }
```

```
            Quasistationary(
         InitialStep= 1e-3 Increment= 1.2
         MinStep= 1e-12 MaxStep= 0.02
         Goal { Name= "G" Voltage=@Vg@ }
       ){ Coupled { Poisson eQuantumPotential Electron Hole } }
         Quasistationary(
         InitialStep= 1e-3 Increment= 1.2
         MinStep= 1e-12 MaxStep= 0.95
         Goal { Name= "D" Voltage=@Vd@ }
               DoZero
          ){ Coupled { Poisson eQuantumPotential Electron Hole } }
         }
       *------------------- END -------------------*
```

3. INSPECT—inspect_inc.cmd

The line of code following # is the prompt character for program designer to take note such that it will not be executed by the computer.

```
#--------------------------------------------------------------------#
#              Script file designed to compute    :           #
#                  * The threshold voltage           :   VT    #
#                  * The transconductance            :   gm    #
#--------------------------------------------------------------------#
if { ! [catch {open n@previous@_ins.log w} log_file] } {
       set fileId stdout
}
puts $log_file " "
puts $log_file "          ------------------------------------ "
puts $log_file "          Values of the extracted Parameters : "
puts $log_file "          ------------------------------------ "
puts $log_file " "
```

```
puts $log_file " "
set   DATE     [ exec   date ]
set   WORK     [ exec pwd    ]
puts $log_file "    Date          : $DATE "
puts $log_file "    Directory : $WORK "
puts $log_file " "
puts $log_file " "
#                                              #
#              idvgs=y(x) ;    vgsvgs=x(x) ;   #
#                                              #
set out_file n@previous@_des
proj_load "${out_file}.plt"
# -------------------------------------------------------------------- #
# I)   VT = Xintercept(maxslope(ID[VGS])) or VT = VGS( IDS= 0.1 ua/um ) #
# -------------------------------------------------------------------- #
cv_create      idvgs   "${out_file} G OuterVoltage" "${out_file} D TotalCurrent"
cv_create      vdsvgs  "${out_file} G OuterVoltage" "${out_file} D OuterVoltage"
#.................................................... #
# 1) VT extracted as the intersection point with the X axis at the point #
#      where the id(vgs) slope reaches its maxmimum :    #
#.................................................... #
set VT1     [ f_VT1 idvgs ]
#...........................................                #
# 2) Printing of the whole set of extracted values (std output) : #
#...........................................                #
puts $log_file "Threshold     voltage VT1     = $VT1   Volts"
puts $log_file " "
# 3) Initialization and display of curves on the main Inspect screen   :   #
cv_display       idvgs
cv_lineStyle   idvgs    solid
cv_lineColor   idvgs    red
# -------------------------------------------------------------------- #
# II)                 gm =   maxslope((ID[VGS])          #
# -------------------------------------------------------------------- #
set gm      [ f_gm idvgs ]
puts $log_file " "
```

```
puts $log_file "Transconductance gm              = $gm   A/V"
puts $log_file " "
set ioff   [ cv_compute "vecmin(<idvgs>)" A A A A ]
puts $log_file " "
puts $log_file "Current ioff          = $ioff   A"
puts $log_file " "
set isat   [ cv_compute "vecmax(<idvgs>)" A A A A ]
puts $log_file " "
puts $log_file "Current isat          = $isat   A"
puts $log_file " "
set rout   [ cv_compute "Rout(<idvgs>)" A A A A ]
puts $log_file " "
puts $log_file "Resistant rout          = $rout   A"
puts $log_file " "
cv_createWithFormula logcurve "log10(<idvgs>)" A A A A
cv_createWithFormula difflog "diff(<logcurve>)" A A A A
set sslop [ cv_compute "1/vecmax(<difflog>)" A A A A ]
puts $log_file " "
puts $log_file "sub solp          = $sslop   A/V"
puts $log_file " "
### Puting into Family Table #####
ft_scalar VT $VT1
ft_scalar gmax $gm
ft_scalar ioff $ioff
ft_scalar isat $isat
ft_scalar sslop $sslop
ft_scalar rout $rout
close $log_file
#----------------- END ---------------#
```

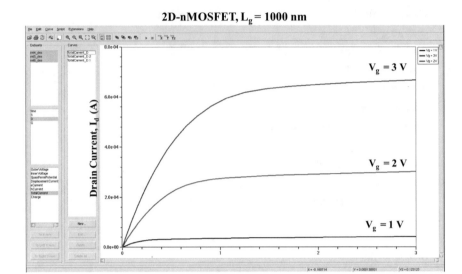

Fig. 2.29 I_d–V_d curve of simulation of 2D nMOSFET in Inspect tool

The electric properties of the simulation result of 2D nMOSFET output curve is as shown in Fig. 2.29 by **Inspect tool**.

2.4 [Example 2.3] 2D p-Type MOSFET with I_d–V_g Characteristics Simulation

The following three main program code files are all based on Synopsys Sentaurus TCAD 2014 version.

This 2D pMOSFET I_d–V_g simulation example is very similar to Example 2.1 nMOSFET I_d–V_g, only have difference in doping and electrodes bias setting. This is the standard example of 2D pMOSFET I_d–V_g example.

1. SDE – devise_dvs.cmd

This is the best example of 2D p MOSFET.

The line of code following; **is the prompt character for program designer to take note such that it will not be executed by the computer**.

```
;------------------ parameter -----------------------;
(define Lg @Lg@)
(define tox @tox@)
(define tac 100)
(define Body 400)
(define LSDC 25)
(define LSD 25)
(define C_Doping 1e16)
(define D_Doping 1e20)
(define S_Doping 1e20)
(define B_Doping 1e15)
(define nm 1e-3)
(define x1 LSDC)
(define x2 (+ x1 LSD))
(define x3 (+ x2 Lg))
(define x4 (+ x3 LSD))
(define x5 (+ x4 LSDC))
(define y1 (- Body))
(define y2 tac)
(define y3 (+ tac tox))
;------------------ Structure -------------------;
"ABA"
;--- source ---
(sdegeo:create-rectangle
   (position     0 0 0)
   (position     x1 y2 0)   "Silicon" "SourceC" )
(sdegeo:create-rectangle
   (position     x1 0 0)
   (position     x2 y2 0)   "Silicon" "Source" )
;--- Channel ---
(sdegeo:create-rectangle
   (position     x2 0 0)
   (position     x3 y2 0)   "Silicon" "Channel" )
;--- Drain ---
(sdegeo:create-rectangle
```

```
   (position      x3 0 0)
   (position      x4 y2 0)   "Silicon" "Drain" )
(sdegeo:create-rectangle
   (position      x4 0 0)
   (position      x5 y2 0)   "Silicon" "DrainC" )
;--- Body ---
(sdegeo:create-rectangle
   (position      0 0 0)
   (position      x5 y1 0)   "Silicon" "Body" )
;--- Gate oxide ---
(sdegeo:create-rectangle
   (position      x2 y2 0)
   (position      x3 y3 0)   "SiO2" "Gateoxide" )
;---------------------- Contact ----------------------;
;----- Gate -----
(sdegeo:define-contact-set "G"
   4.0   (color:rgb 1.0 0.0 0.0 ) "##")
(sdegeo:define-2d-contact   (find-edge-id (position (+ x2 (/ Lg 2)) y3 0)) "G")
;----- Source -----
(sdegeo:define-contact-set "S"
   4.0   (color:rgb 1.0 0.0 0.0 ) "##")
(sdegeo:define-2d-contact   (find-edge-id (position 10 tac 0)) "S")
;----- Drain -----
(sdegeo:define-contact-set "D"
   4.0   (color:rgb 1.0 0.0 0.0 ) "##")
(sdegeo:define-2d-contact   (find-edge-id (position (+ 50 Lg 35) tac 0)) "D")
;----- Substrate -----
(sdegeo:define-contact-set "substrate"
   4.0   (color:rgb 1.0 0.0 0.0 ) "##")
(sdegeo:define-2d-contact   (find-edge-id (position (+ x2 (/ Lg 2)) (- Body) 0))
"substrate")
;--------------------- Doping -----------------------;
;--- Channel ---
(sdedr:define-constant-profile "dopedC" "PhosphorusActiveConcentration"
C_Doping )
```

```
(sdedr:define-constant-profile-region    "RegionC" "dopedC" "Channel" )
;--- Source ---
(sdedr:define-constant-profile "dopedS" "BoronActiveConcentration" S_Doping )
(sdedr:define-constant-profile-region    "RegionS" "dopedS" "Source" )
(sdedr:define-constant-profile "dopedSC" "BoronActiveConcentration" S_Doping )
(sdedr:define-constant-profile-region    "RegionSC" "dopedSC" "SourceC" )
;--- Drain ---
(sdedr:define-constant-profile "dopedD" "BoronActiveConcentration" D_Doping )
(sdedr:define-constant-profile-region    "RegionD" "dopedD" "Drain" )
(sdedr:define-constant-profile "dopedDC" "BoronActiveConcentration" D_Doping )
(sdedr:define-constant-profile-region    "RegionDC" "dopedDC" "DrainC" )
;--- Body ---
(sdedr:define-constant-profile "dopedB" "PhosphorusActiveConcentration"
B_Doping )
(sdedr:define-constant-profile-region    "RegionB" "dopedB" "Body" )
;---------------------- Mesh -----------------------;
;--- AllMesh ---
(sdedr:define-refinement-size "Cha_Mesh" 20 20 0 10 10 0)
(sdedr:define-refinement-material "channel_RF" "Cha_Mesh" "Silicon" )
;--- ChannelMesh ---
(sdedr:define-refinement-window "multiboxChannel" "Rectangle"
(position 25 (- 50) 0)
(position (+ 50 Lg 25) (+ tac 50) 0))
(sdedr:define-multibox-size "multiboxSizeChannel"   5 5 0 1 1 0)
(sdedr:define-multibox-placement "multiboxPlacementChannel"
"multiboxSizeChannel" "multiboxChannel")
(sdedr:define-refinement-function "multiboxPlacementChannel"
"DopingConcentration" "MaxTransDiff" 1)
;---------------- Save BND and CMD and rescale to nm ----------------------;
(sde:assign-material-and-region-names (get-body-list) )
(sdeio:save-tdr-bnd (get-body-list) "n@node@_nm.tdr")
(sdedr:write-scaled-cmd-file "n@node@_msh.cmd" nm)
(define sde:scale-tdr-bnd
  (lambda (tdrin sf tdrout)
    (sde:clear)
```

```
        (sdegeo:set-default-boolean "XX")
        (sdeio:read-tdr-bnd tdrin)
        (entity:scale (get-body-list) sf)
        (sdeio:save-tdr-bnd (get-body-list) tdrout)
        )
    )
(sde:scale-tdr-bnd "n@node@_nm.tdr" nm "n@node@_bnd.tdr")
;---------------- END ----------------;
```

2. SDVICE – dessis_des.cmd

The line of code following *** and # are the prompt characters** for program designer to take note such that it will not be executed by the computer.

```
Electrode{
{name="D" voltage=0.0}
{name="S" voltage=0.0}
{name="G" voltage=0.0 WorkFunction=@WK@}
}
File{
        Grid="@tdr@"
        Plot="@tdrdat@"
        Current="@plot@"
        Output="@log@"
        parameter="@parameter@"
}
Physics{
    Mobility( DopingDep HighFieldSaturation Enormal )
    EffectiveIntrinsicDensity( OldSlotboom )
```

```
    Recombination( SRH(DopingDep) )
    hQuantumPotential
}
Math{
    -CheckUndefinedModels
    Number_Of_Threads=4
    Extrapolate
    Derivatives
    * Avalderivatives
    RelErrControl
    Digits=5
    ErRef(electron)=1.e10
    ErRef(hole)=1.e10
    Notdamped=50
    Iterations=20
    Directcurrent
    Method=ParDiSo
    Parallel= 2
*-VoronoiFaceBoxMethod
    NaturalBoxMethod
}
Plot{
    eDensity hDensity
    eCurrent hCurrent
    TotalCurrent/Vector eCurrent/Vector hCurrent/Vector
    eMobility hMobility
    eVelocity hVelocity
    eEnormal hEnormal
    ElectricField/Vector Potential SpaceCharge
    eQuasiFermi hQuasiFermi
    Potential Doping SpaceCharge
    SRH Auger
    AvalancheGeneration
```

```
    DonorConcentration AcceptorConcentration
    Doping
    eGradQuasiFermi/Vector hGradQuasiFermi/Vector
    eEparallel hEparalllel
    BandGap
    BandGapNarrowing
    Affinity
    ConductionBand ValenceBand
    hQuantumPotential
 }
 Solve {
   Coupled ( Iterations= 150){ Poisson hQuantumPotential }
   Coupled { Poisson hQuantumPotential Electron Hole }
   Quasistationary(
      InitialStep= 1e-3 Increment= 1.2
      MinStep= 1e-12 MaxStep= 0.95
      Goal { Name= "D" Voltage=@Vd@ }
   ){ Coupled { Poisson hQuantumPotential Electron Hole } }
      Quasistationary(
      InitialStep= 1e-3 Increment= 1.2
      MinStep= 1e-12 MaxStep= 0.02
      Goal { Name= "G" Voltage=@Vg@ }
      DoZero
   ){ Coupled { Poisson hQuantumPotential Electron Hole } }
 }
 *-------------- END ---------------*
```

3. INSPECT – inspect_inc.cmd

The line of code following # **is the prompt character for program designer to take note such that it will not be executed by the computer**.

```
#---------------------------------------------------------------------#
#          Script file designed to compute    :        #
#             * The threshold voltage             :    VT
#
#             * The transconductance              :    gm    #
#---------------------------------------------------------------------#
if { ! [catch {open n@previous@_ins.log w} log_file] } {
     set fileId stdout
}
puts $log_file " "
puts $log_file "            ----------------------------------- "
puts $log_file "            Values of the extracted Parameters : "
puts $log_file "            ----------------------------------- "
puts $log_file " "
puts $log_file " "
set   DATE    [ exec   date ]
set   WORK    [ exec pwd    ]
puts $log_file "    Date      : $DATE "
puts $log_file "    Directory : $WORK "
puts $log_file " "
puts $log_file " "
#                                              #
#                            idvgs=y(x) ;    vgsvgs=x(x) ; #
#set out_file n@previous@_des
proj_load "${out_file}.plt"
# ------------------------------------------------------------------- #
# I) VT = Xintercept(maxslope(ID[VGS])) or VT = VGS( IDS= 100nA/um ) #
# ------------------------------------------------------------------- #
cv_create     idvgs    "${out_file} G OuterVoltage" "${out_file} S TotalCurrent"
cv_create     vdsvgs   "${out_file} G OuterVoltage" "${out_file} S OuterVoltage"
#.................................................. #
# 1) VT extracted as the intersection point with the X axis at the point #
#     where the id(vgs) slope reaches its maxmimum :              #
#.................................................. #
set VT1    [ f_VT1 idvgs ]
```

```
#............................................................           #
# 2) Printing of the whole set of extracted values (std output) :         #
#............................................................           #
puts $log_file "Threshold     voltage VT1    = $VT1    Volts"
puts $log_file " "
#...............................................................   #
# 3) Initialization and display of curves on the main Inspect screen   :   #
cv_display      idvgs
cv_lineStyle   idvgs   solid
cv_lineColor   idvgs   red
# ----------------------------------------------------------------------- #
# II)                  gm =   maxslope((ID[VGS])                  #
# ----------------------------------------------------------------------- #
set gm        [ f_gm idvgs ]
puts $log_file " "
puts $log_file "Transconductance gm    = $gm   A/V"
puts $log_file " "
set ioff   [ cv_compute "vecmin(<idvgs>)" A A A A ]
puts $log_file " "
puts $log_file "Current ioff             = $ioff   A"
puts $log_file " "
set isat   [ cv_compute "vecmax(<idvgs>)" A A A A ]
puts $log_file " "
puts $log_file "Current isat             = $isat   A"
puts $log_file " "
set rout   [ cv_compute "Rout(<idvgs>)" A A A A ]
puts $log_file " "
puts $log_file "Resistant rout           = $rout   A"
puts $log_file " "
cv_createWithFormula logcurve "log10(<idvgs>)" A A A A
cv_createWithFormula difflog "(-1)*diff(<logcurve>)" A A A A
set sslop [ cv_compute "1/vecmax(<difflog>)" A A A A ]
puts $log_file " "
puts $log_file "sub solp             = $sslop   A/V"
puts $log_file " "
```

Puting into Family Table

ft_scalar VT $VT1

ft_scalar gmax $gm

ft_scalar ioff $ioff

ft_scalar isat $isat

ft_scalar sslop $sslop

ft_scalar rout $rout

close $log_file

#----------------------- END -----------------------#

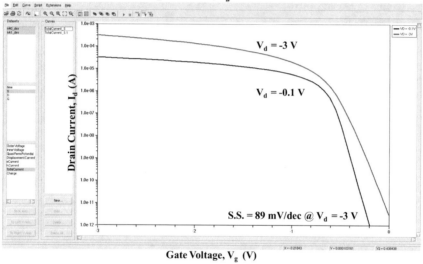

Fig. 2.30 I_d–V_g curve of simulation of 2D p-type MOSFET

The electric property of 2D pMOSFET transfer curve is as shown in Fig. 2.30.

2.5 [Example 2.4] 2D p-Type MOSFET with I_d–V_g Characteristics Simulation

This 2D pMOSFET I_d–V_d simulation example is very similar to Example 2.3 nMOSFET, only have difference in doping and electrodes bias setting. This is the standard example of 2D pMOSFET I_d–V_g.

The following three main program code files are all based on Synopsys Sentaurus TCAD 2014 version.

1. SDE – devise_dvs.cmd

The line of code following ; **is the prompt character for program designer to take note such that it will not be executed by the computer**.

```
;-------------------- parameter -----------------------------;
(define Lg @Lg@)
(define tox @tox@)
(define tac 100)
(define Body 400)
(define LSDC 25)
(define LSD 25)
(define C_Doping 1e16)
(define D_Doping 1e20)
(define S_Doping 1e20)
(define B_Doping 1e15)
(define nm 1e-3)
(define x1 LSDC)
(define x2 (+ x1 LSD))
(define x3 (+ x2 Lg))
(define x4 (+ x3 LSD))
(define x5 (+ x4 LSDC))
(define y1 (- Body))
(define y2 tac)
(define y3 (+ tac tox))
;---------------------- Structure ----------------------;
"ABA"
;--- source ---
(sdegeo:create-rectangle
   (position      0 0 0)
   (position      x1 y2 0)   "Silicon" "SourceC" )
(sdegeo:create-rectangle
   (position      x1 0 0)
   (position      x2 y2 0)   "Silicon" "Source" )
;--- Channel ---
(sdegeo:create-rectangle
   (position      x2 0 0)
   (position      x3 y2 0)   "Silicon" "Channel" )
;--- Drain ---
(sdegeo:create-rectangle
```

```
(position       x3 0 0)
(position       x4 y2 0)    "Silicon" "Drain" )
(sdegeo:create-rectangle
(position       x4 0 0)
(position       x5 y2 0)    "Silicon" "DrainC" )
;--- Body ---
(sdegeo:create-rectangle
(position       0 0 0)
(position       x5 y1 0)    "Silicon" "Body" )
;--- Gate oxide ---
(sdegeo:create-rectangle
(position       x2 y2 0)
(position       x3 y3 0)    "SiO2" "Gateoxide" )
;----------------------- Contact -----------------------;
;----- Gate -----
(sdegeo:define-contact-set "G"
   4.0    (color:rgb 1.0 0.0 0.0 ) "##")
(sdegeo:define-2d-contact    (find-edge-id (position (+ x2 (/ Lg 2)) y3 0)) "G")
;----- Source -----
(sdegeo:define-contact-set "S"
   4.0    (color:rgb 1.0 0.0 0.0 ) "##")
(sdegeo:define-2d-contact    (find-edge-id (position 10 tac 0)) "S")
;----- Drain -----
(sdegeo:define-contact-set "D"
   4.0    (color:rgb 1.0 0.0 0.0 ) "##")
(sdegeo:define-2d-contact    (find-edge-id (position (+ 50 Lg 35) tac 0)) "D")
;----- Substrate -----
(sdegeo:define-contact-set "substrate"
   4.0    (color:rgb 1.0 0.0 0.0 ) "##")
(sdegeo:define-2d-contact    (find-edge-id (position (+ x2 (/ Lg 2)) (- Body) 0))
"substrate")
;-------------------------- Doping ---------------------------;
;--- Channel ---
(sdedr:define-constant-profile "dopedC" "PhosphorusActiveConcentration"
C_Doping )
```

```
(sdedr:define-constant-profile-region    "RegionC" "dopedC" "Channel" )
;--- Source ---
(sdedr:define-constant-profile "dopedS" "BoronActiveConcentration" S_Doping )
(sdedr:define-constant-profile-region    "RegionS" "dopedS" "Source" )
(sdedr:define-constant-profile "dopedSC" "BoronActiveConcentration" S_Doping )
(sdedr:define-constant-profile-region    "RegionSC" "dopedSC" "SourceC" )
;--- Drain ---
(sdedr:define-constant-profile "dopedD" "BoronActiveConcentration" D_Doping )
(sdedr:define-constant-profile-region    "RegionD" "dopedD" "Drain" )
(sdedr:define-constant-profile "dopedDC" "BoronActiveConcentration" D_Doping )
(sdedr:define-constant-profile-region    "RegionDC" "dopedDC" "DrainC" )
;--- Body ---
(sdedr:define-constant-profile "dopedB" "PhosphorusActiveConcentration"
B_Doping )
(sdedr:define-constant-profile-region    "RegionB" "dopedB" "Body" )
;-------------------------- Mesh ------------------------------------;
;--- AllMesh ---
(sdedr:define-refinement-size "Cha_Mesh" 20 20 0 10 10 0)
(sdedr:define-refinement-material "channel_RF" "Cha_Mesh" "Silicon" )
;--- ChannelMesh ---
(sdedr:define-refinement-window "multiboxChannel" "Rectangle"
(position 25 (- 50) 0)
(position (+ 50 Lg 25) (+ tac 50) 0))
(sdedr:define-multibox-size "multiboxSizeChannel"    5 5 0 1 1 0)
(sdedr:define-multibox-placement "multiboxPlacementChannel"
"multiboxSizeChannel" "multiboxChannel")
(sdedr:define-refinement-function "multiboxPlacementChannel"
"DopingConcentration" "MaxTransDiff" 1)
;--------------- Save BND and CMD and rescale to nm -------------;
(sde:assign-material-and-region-names (get-body-list) )
(sdeio:save-tdr-bnd (get-body-list) "n@node@_nm.tdr")
(sdedr:write-scaled-cmd-file "n@node@_msh.cmd" nm)
(define sde:scale-tdr-bnd
  (lambda (tdrin sf tdrout)
    (sde:clear)
```

```
(sdegeo:set-default-boolean "XX")
(sdeio:read-tdr-bnd tdrin)
(entity:scale (get-body-list) sf)
(sdeio:save-tdr-bnd (get-body-list) tdrout)
)
)
(sde:scale-tdr-bnd "n@node@_nm.tdr" nm "n@node@_bnd.tdr")
;---------------------------- END -------------------------------;
```

2. SDVICE – dessis_des.cmd

The line of code following **# and * are the prompt characters** for program designer to take note such that it will not be executed by the computer.

```
Electrode{
{name="D" voltage=0.0}
{name="S" voltage=0.0}
{name="G" voltage=0.0 WorkFunction=@WK@}
}
File{
        Grid="@tdr@"
        Plot="@tdrdat@"
        Current="@plot@"
        Output="@log@"
        parameter="@parameter@"
}
Physics{
     Mobility( DopingDep HighFieldSaturation Enormal )
```

```
      EffectiveIntrinsicDensity( OldSlotboom )
      Recombination( SRH(DopingDep) )
      hQuantumPotential
}
Math{
    -CheckUndefinedModels
    Number_Of_Threads=4
    Extrapolate
    Derivatives
    * Avalderivatives
    RelErrControl
    Digits=5
    ErRef(electron)=1.e10
    ErRef(hole)=1.e10
    Notdamped=50
    Iterations=20
    Directcurrent
    Method=ParDiSo
    Parallel= 2
*-VoronoiFaceBoxMethod
    NaturalBoxMethod
}
Plot{
  eDensity hDensity
  eCurrent hCurrent
  TotalCurrent/Vector eCurrent/Vector hCurrent/Vector
  eMobility hMobility
  eVelocity hVelocity
  eEnormal hEnormal
  ElectricField/Vector Potential SpaceCharge
  eQuasiFermi hQuasiFermi
  Potential Doping SpaceCharge
  SRH Auger
```

```
    AvalancheGeneration
    DonorConcentration AcceptorConcentration
    Doping
     eGradQuasiFermi/Vector hGradQuasiFermi/Vector
     eEparallel hEparalllel
     BandGap
     BandGapNarrowing
     Affinity
     ConductionBand ValenceBand
     hQuantumPotential
  }
  Solve {
     Coupled ( Iterations= 150){ Poisson hQuantumPotential }
     Coupled { Poisson hQuantumPotential Electron Hole }
          Quasistationary(
        InitialStep= 1e-3 Increment= 1.2
        MinStep= 1e-12 MaxStep= 0.02
        Goal { Name= "G" Voltage=@Vg@ }
      ){ Coupled { Poisson hQuantumPotential Electron Hole } }
        Quasistationary(
        InitialStep= 1e-3 Increment= 1.2
        MinStep= 1e-12 MaxStep= 0.95
        Goal { Name= "D" Voltage=@Vd@ }
        DoZero
      ){ Coupled { Poisson hQuantumPotential Electron Hole } }
   }
  *------------------------ END ------------------------*
```

3. INSPECT – inspect_inc.cmd

The line of code following **# is the prompt character** for program designer to take note such that it will not be executed by the computer.

```
#------------------------------------------------------------------------#
#            Script file designed to compute    :              #
#               * The threshold voltage                :   VT
#               * The transconductance                 :   gm   #
#------------------------------------------------------------------------#
if { ! [catch {open n@previous@_ins.log w} log_file] } {
        set fileId stdout
}
puts $log_file " "
puts $log_file "              ----------------------------------- "
puts $log_file "              Values of the extracted Parameters : "
puts $log_file "              ----------------------------------- "
puts $log_file " "
puts $log_file " "
set    DATE     [ exec   date ]
set    WORK     [ exec pwd    ]
puts $log_file "    Date       : $DATE "
puts $log_file "    Directory : $WORK "
puts $log_file " "
puts $log_file " "
#                                         #
#                        idvgs=y(x) ;   vgsvgs=x(x) ;        #
set out_file n@previous@_des
```

```
proj_load "${out_file}.plt"
# I)   VT = Xintercept(maxslope(ID[VGS])) or VT = VGS( IDS= 0.1 ua/um ) #
# -------------------------------------------------------------------- #
cv_create     idvgs     "${out_file} G OuterVoltage" "${out_file} S TotalCurrent"
cv_create     vdsvgs    "${out_file} G OuterVoltage" "${out_file} S OuterVoltage"
#................................................... #
# 1) VT extracted as the intersection point with the X axis at the point #
#     where the id(vgs) slope reaches its maxmimum :           #
#................................................. #
set VT1     [ f_VT1 idvgs ]
#.............................................        #
# 2) Printing of the whole set of extracted values (std output) :     #
#.............................................        #
puts $log_file "Threshold     voltage VT1   = $VT1   Volts"
puts $log_file " "
#.............................................   #
# 3) Initialization and display of curves on the main Inspect screen   :  #
# .............................................   #
cv_display      idvgs
cv_lineStyle   idvgs   solid
cv_lineColor   idvgs   red
# -------------------------------------------------------------------- #
# II)                   gm =  maxslope((ID[VGS])            #
# -------------------------------------------------------------------- #
set gm      [ f_gm idvgs ]
puts $log_file " "
puts $log_file "Transconductance gm             = $gm   A/V"
puts $log_file " "
set ioff   [ cv_compute "vecmin(<idvgs>)" A A A A ]
```

```
puts $log_file " "
puts $log_file "Current ioff              = $ioff    A"
puts $log_file " "
set isat   [ cv_compute "vecmax(<idvgs>)" A A A A ]
puts $log_file " "
puts $log_file "Current isat              = $isat    A"
puts $log_file " "
set rout   [ cv_compute "Rout(<idvgs>)" A A A A ]
puts $log_file " "
puts $log_file "Resistant rout            = $rout    A"
puts $log_file " "
cv_createWithFormula logcurve "log10(<idvgs>)" A A A A
cv_createWithFormula difflog "(-1)*diff(<logcurve>)" A A A A
set sslop [ cv_compute "1/vecmax(<difflog>)" A A A A ]
puts $log_file " "
puts $log_file "sub solp              = $sslop    A/V"
puts $log_file " "
### Puting into Family Table #####
ft_scalar VT $VT1
ft_scalar gmax $gm
ft_scalar ioff $ioff
ft_scalar isat $isat
ft_scalar sslop $sslop
ft_scalar rout $rout
close $log_file
#--------------------------- END ---------------------------#
```
The electric property of 2D pMOSFET I_s–V_d is as shown in Fig. 2.31.

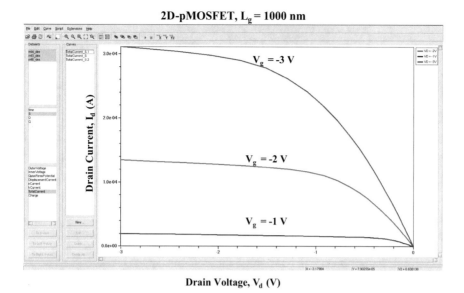

Fig. 2.31 I_s–V_d curve of the simulation of 2D pMOSFET by Inspect tool

2.6 [Example 2.5] 2D n-Type MOSFET with LDD (Lightly Doped Drain) Simulation

This is the standard example of 2D LDD nMOSFET. This 2D LDD nMOSFET I_d–V_g simulation example is very similar to Example 2.1 nMOSFET, only add and LDD doping region.

The following three main program code files are all based on Synopsys Sentaurus TCAD 2014 version.

Description: LDD (lightly doped drain) is an extremely effective method for reducing SCE of 2D MOSFET. The breakdown voltage at the junction is the function of highest electric field. When channel length is reduced, the bias voltage might not be reduced by the same ratio, such that the junction electric field will get even higher, which will make the effects of approximating accumulated breakdown and approximating penetration become more significant. In addition, when the device dimension is reduced, the parasite BJT will become more decisive, and the breakdown effect will be enhanced [1].

A method for reducing these breakdown effects is to change the dopant distribution of drain contact. By using region with light doping, the peak electric field in

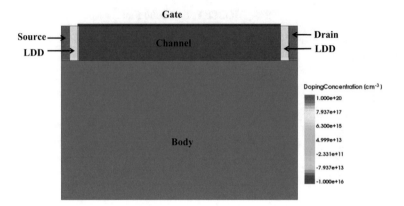

Fig. 2.32 Simulation of LDD device structure of 2D nMOSFET

the spatial charge region will be reduced, thus minimizing the breakdown effect. As for the peak value at drain junction, the electric field is the function of semiconductor doping and the function of curvature of n^+ drain region. In the LDD structure, the electric field of oxide-semiconductor junction is lower than the traditional structure. Among traditional devices, electric fields usually peak at the metallurgical junction, and it will be quickly reduced to zero at the drain. This is because the electric field cannot exist in the highly conductive n^+ region. On the other hand, the electric field in LDD device will be extended across the n region before being reduced to zero, and this effect will minimize the breakdown effect and hot carrier effect.

There are two disadvantages of LDD device. For one, the fabrication complexity is increased. For the other, the drain resistance is increased. Nonetheless, this extra process step can indeed fabricate the device with significantly improved performance. The cross section of LDD device is as shown in Fig. 2.32, in which the source terminal is changed to the lightly doped n region, which will lead to improve the device operating performance while reducing the process complexity. The series resistance will lead to increased device power consumption, so this factor must be taken into consideration for high-power device.

1. SDE – devise_dvs.cmd

The line of code following ; **is the prompt character for program designer to take note such that it will not be executed by the computer**;

```
;----------------------- parameter ----------------------------;
(define Lg @Lg@)
(define tox @tox@)
(define tac 100)
(define Body 400)
(define LSDC 25)
(define LSD 25)
(define C_Doping 1e16)
(define DC_Doping 1e20)
(define D_Doping 1e18)
(define S_Doping 1e18)
(define SC_Doping 1e20)
;(define B_Doping 1e15)
(define B_Doping 1e16)
(define nm 1e-3)
(define x1 LSDC)
(define x2 (+ x1 LSD))
(define x3 (+ x2 Lg))
(define x4 (+ x3 LSD))
(define x5 (+ x4 LSDC))
(define y1 (- Body))
(define y2 tac)
(define y3 (+ tac tox))
;------------------------ Structure ------------------------;
"ABA"
;--- source ---
(sdegeo:create-rectangle
   (position      0 0 0)
   (position      x1 y2 0)   "Silicon" "SourceC" )
(sdegeo:create-rectangle
   (position      x1 0 0)
   (position      x2 y2 0)   "Silicon" "Source" )
;--- Channel ---
(sdegeo:create-rectangle
   (position      x2 0 0)
```

```
    (position      x3 y2 0)    "Silicon" "Channel" )
;--- Drain ---
(sdegeo:create-rectangle
    (position      x3 0 0)
    (position      x4 y2 0)    "Silicon" "Drain" )
(sdegeo:create-rectangle
    (position      x4 0 0)
    (position      x5 y2 0)    "Silicon" "DrainC" )
;--- Body ---
(sdegeo:create-rectangle
    (position      0 0 0)
    (position      x5 y1 0)    "Silicon" "Body" )
;--- Gate oxide ---
(sdegeo:create-rectangle
    (position      x2 y2 0)
    (position      x3 y3 0)    "SiO2" "Gateoxide" )
;--------------------------- Contact ---------------------------;
;----- Gate -----
(sdegeo:define-contact-set "G"
    4.0   (color:rgb 1.0 0.0 0.0 ) "##")
(sdegeo:define-2d-contact   (find-edge-id (position (+ x2 (/ Lg 2)) y3 0)) "G")
;----- Source -----
(sdegeo:define-contact-set "S"
    4.0   (color:rgb 1.0 0.0 0.0 ) "##")
(sdegeo:define-2d-contact   (find-edge-id (position 10 tac 0)) "S")
;----- Drain -----
(sdegeo:define-contact-set "D"
    4.0   (color:rgb 1.0 0.0 0.0 ) "##")
(sdegeo:define-2d-contact   (find-edge-id (position (+ 50 Lg 35) tac 0)) "D")
;----- Substrate -----
(sdegeo:define-contact-set "substrate"
    4.0   (color:rgb 1.0 0.0 0.0 ) "##")
(sdegeo:define-2d-contact   (find-edge-id (position (+ x2 (/ Lg 2)) (- Body) 0))
"substrate")
;------------------------- Doping -----------------------------;
```

```
;--- Channel ---
(sdedr:define-constant-profile "dopedC" "BoronActiveConcentration" C_Doping )
(sdedr:define-constant-profile-region    "RegionC" "dopedC" "Channel" )
;--- Source ---
(sdedr:define-constant-profile "dopedS" "PhosphorusActiveConcentration"
S_Doping )
(sdedr:define-constant-profile-region    "RegionS" "dopedS" "Source" )
(sdedr:define-constant-profile "dopedSC" "PhosphorusActiveConcentration"
SC_Doping )
(sdedr:define-constant-profile-region    "RegionSC" "dopedSC" "SourceC" )
;--- Drain ---
(sdedr:define-constant-profile "dopedD" "PhosphorusActiveConcentration"
D_Doping )
(sdedr:define-constant-profile-region    "RegionD" "dopedD" "Drain" )
(sdedr:define-constant-profile "dopedDC" "PhosphorusActiveConcentration"
DC_Doping )
(sdedr:define-constant-profile-region    "RegionDC" "dopedDC" "DrainC" )
;--- Body ---
(sdedr:define-constant-profile "dopedB" "BoronActiveConcentration" B_Doping )
(sdedr:define-constant-profile-region    "RegionB" "dopedB" "Body" )
;--------------------------- Mesh -----------------------------------;
;--- AllMesh ---
(sdedr:define-refinement-size "Cha_Mesh" 20 20 0 10 10 0)
(sdedr:define-refinement-material "channel_RF" "Cha_Mesh" "Silicon" )
;--- ChannelMesh ---
(sdedr:define-refinement-window "multiboxChannel" "Rectangle"
(position 25 (- 50) 0)
(position (+ 50 Lg 25) (+ tac 50) 0))
(sdedr:define-multibox-size "multiboxSizeChannel"    5 5 0 1 1 0)
(sdedr:define-multibox-placement "multiboxPlacementChannel"
"multiboxSizeChannel" "multiboxChannel")
(sdedr:define-refinement-function "multiboxPlacementChannel"
"DopingConcentration" "MaxTransDiff" 1)
;---------------------- Save BND and CMD and rescale to nm ---------------------------;
(sde:assign-material-and-region-names (get-body-list) )
```

```
(sdeio:save-tdr-bnd (get-body-list) "n@node@_nm.tdr")
(sdedr:write-scaled-cmd-file "n@node@_msh.cmd" nm)
(define sde:scale-tdr-bnd
  (lambda (tdrin sf tdrout)
    (sde:clear)
    (sdegeo:set-default-boolean "XX")
    (sdeio:read-tdr-bnd tdrin)
    (entity:scale (get-body-list) sf)
    (sdeio:save-tdr-bnd (get-body-list) tdrout)
    )
  )
(sde:scale-tdr-bnd "n@node@_nm.tdr" nm "n@node@_bnd.tdr")
;------------------------------------- END -------------------------------------;
```

2. SDVICE – dessis_des.cmd

The line of code following **# and * are the prompt characters** for program designer to take note such that it will not be executed by the computer.

```
Electrode{
{name="D" voltage=0.0}
{name="S" voltage=0.0}
{name="G" voltage=0.0 WorkFunction=@WK@}
}
File{
      Grid="@tdr@"
      Plot="@tdrdat@"
      Current="@plot@"
      Output="@log@"
      parameter="@parameter@"
}
Physics{
    Mobility( DopingDep HighFieldSaturation Enormal )
    EffectiveIntrinsicDensity( OldSlotboom )
    Recombination( SRH(DopingDep) )
    eQuantumPotential
}
Math{
    -CheckUndefinedModels
```

```
        Number_Of_Threads=4
        Extrapolate
        Derivatives
        * Avalderivatives
        RelErrControl
        Digits=5
        ErRef(electron)=1.e10
        ErRef(hole)=1.e10
        Notdamped=50
        Iterations=20
        Directcurrent
        Method=ParDiSo
        Parallel= 2
    *-VoronoiFaceBoxMethod
        NaturalBoxMethod
    }
    Plot{
        eDensity hDensity
        eCurrent hCurrent
        TotalCurrent/Vector eCurrent/Vector hCurrent/Vector
        eMobility hMobility
        eVelocity hVelocity
        eEnormal hEnormal
        ElectricField/Vector Potential SpaceCharge
        eQuasiFermi hQuasiFermi
        Potential Doping SpaceCharge
        SRH Auger
        AvalancheGeneration
        DonorConcentration AcceptorConcentration


        Doping
        eGradQuasiFermi/Vector hGradQuasiFermi/Vector
        eEparallel hEparalllel
        BandGap
        BandGapNarrowing
```

```
        Affinity
        ConductionBand ValenceBand
        eQuantumPotential
    }
    Solve {
      Coupled ( Iterations= 150){ Poisson eQuantumPotential }
      Coupled { Poisson eQuantumPotential Electron Hole }
      Quasistationary(
         InitialStep= 1e-3 Increment= 1.2
         MinStep= 1e-12 MaxStep= 0.95
         Goal { Name= "D" Voltage=@Vd@ }
      ){ Coupled { Poisson eQuantumPotential Electron Hole } }
         Quasistationary(
         InitialStep= 1e-3 Increment= 1.2
         MinStep= 1e-12 MaxStep= 0.02
         Goal { Name= "G" Voltage=@Vg@ }
         DoZero
      ){ Coupled { Poisson eQuantumPotential Electron Hole } }
      }
      *-------------------------------------- END ----------------------------------------*
```

3. INSPECT – inspect_inc.cmd

The line of code following # **is the prompt character** for program designer to take note such that it will not be executed by the computer.

```
#----------------------------------------------------------------------#
#           Script file designed to compute    :           #
#               * The threshold voltage   :    VT#          #
#               * The transconductance           :  gm     #
#----------------------------------------------------------------------#
if { ! [catch {open n@previous@_ins.log w} log_file] } {
     set fileId stdout
}
puts $log_file " "
puts $log_file "           ------------------------------------ "
puts $log_file "             Values of the extracted Parameters : "
puts $log_file "           ------------------------------------ "
puts $log_file " "
puts $log_file " "
set   DATE    [ exec   date ]
set   WORK    [ exec pwd    ]
puts $log_file "     Date      : $DATE "
puts $log_file "     Directory : $WORK "
puts $log_file " "
puts $log_file " "
#                                           #
#               idvgs=y(x) ;    vgsvgs=x(x) ; #
set out_file n@previous@_des
proj_load "${out_file}.plt"
# --------------------------------------------------------------------- #
# I)   VT = Xintercept(maxslope(ID[VGS]))   or    VT = VGS( IDS= 0.1 ua/um ) #
# --------------------------------------------------------------------- #
cv_create    idvgs    "${out_file} G OuterVoltage" "${out_file} D TotalCurrent"
cv_create    vdsvgs   "${out_file} G OuterVoltage" "${out_file} D OuterVoltage"
#.................................................. #
# 1) VT extracted as the intersection point with the X axis at the point #
#     where the id(vgs) slope reaches its maxmimum :           #
#.................................................. #
set VT1    [ f_VT1 idvgs ]
#...........................................           #
```

```
# 2) Printing of the whole set of extracted values (std output) :          #
#.........................................................          #
puts $log_file "Threshold     voltage VT1     = $VT1   Volts"
puts $log_file " "
#..................................................................   #
# 3) Initialization and display of curves on the main Inspect screen   :   #
# ...................................................................   #
cv_display      idvgs
cv_lineStyle   idvgs    solid
cv_lineColor   idvgs    red
# --------------------------------------------------------------------- #
# II)                    gm =   maxslope((ID[VGS])                #
# --------------------------------------------------------------------- #
set gm       [ f_gm idvgs ]
puts $log_file " "
puts $log_file "Transconductance gm                 = $gm    A/V"
puts $log_file " "
set ioff   [ cv_compute "vecmin(<idvgs>)" A A A A ]
puts $log_file " "
puts $log_file "Current ioff          = $ioff   A"
puts $log_file " "
set isat   [ cv_compute "vecmax(<idvgs>)" A A A A ]
puts $log_file " "
puts $log_file "Current isat          = $isat   A"
puts $log_file " "
set rout   [ cv_compute "Rout(<idvgs>)" A A A A ]
puts $log_file " "
puts $log_file "Resistant rout          = $rout   A"
puts $log_file " "
cv_createWithFormula logcurve "log10(<idvgs>)" A A A A
cv_createWithFormula difflog "diff(<logcurve>)" A A A A
set sslop [ cv_compute "1/vecmax(<difflog>)" A A A A ]
puts $log_file " "
puts $log_file "sub solp          = $sslop   A/V"
puts $log_file " "
```

Puting into Family Table
ft_scalar VT $VT1
ft_scalar gmax $gm
ft_scalar ioff $ioff
ft_scalar isat $isat
ft_scalar sslop $sslop
ft_scalar rout $rout
close $log_file
#----------------------------------- END ---------------------------------#

The electric property is as shown in Fig. 2.33, which shows that when LDD (lightly doped drain) is added into the original device, the leakage current (I_{off}) is reduced and the sub-threshold slope (SS) is significantly improved as compared to Fig. 2.23. Using LDD for reducing short-channel effect, the 2D n-type MOSFET still has good performance at L_g = 400 nm.

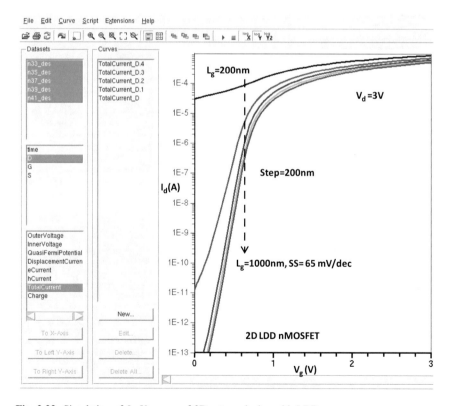

Fig. 2.33 Simulation of I_d–V_g curve of 2D n-type device with LDD

The more complicated well-doping process, anti-punch through process, and retrograde process can be simulated and developed by readers.

2.7 Summary

We introduce fundamental applications of TCAD simulation software, including electric property and physical property analysis in several 2D MOSFET simulations. The meanings of various tool codes are also explained. Readers can get a quick start for using Synopsys Sentaurus TCAD 2014 version.

References

1. S.M. Sze, K.K. Ng, Physics of Semiconductor Devices, 3rd ed (Wiley, New York, 2007)
2. TCAD Sentaurus Device, Synopsys SDevice Ver. J-2014.09 (Synopsys, Inc., Mountain View, CA, USA, 2014)
3. C.C. Hu, Modern Semiconductor Devices for Integrated Circuits (Pearson Education, Inc., 2010)

Chapter 3
3D FinFET with L_g = 15 nm and L_g = 10 nm Simulation

3.1 Introduction of FinFET

In 1965, Gordon Moore proposed the rule that the number of devices on the wafer would be doubled every 18–24 months. This "**Moore's law**" describes the continuous and rapid trend of scaling. Every reduction of feature size will be called a technology generation or technology node. The technology nodes include 0.18, 0.13, 90, 65, and 45 μm. These numbers represent the minimum feature size. For every new **technology node**, all circuit layout properties (such as SRAM cell and CPU) will have their size reduced to 70%. From the historical perspective, new technology node will be generated once every 2 or 3 years. The advantage of new technology node is that the circuit size can be reduced in half of original feature length (70%) and width means 50% reduction of area rate ($0.7 \times 0.7 = 0.49$). For every new technology node, there can be twice as many circuits on one wafer such that the cost of every circuit will be greatly reduced, thus reducing the cost of IC. Moreover, by scaling, the device performances will also enhance such as IC operation speed increasing and power reduction [1].

In addition to gate length (L_g) and width W scaling down, many other parameters will be reduced along with the scaling rule. For example, the effective thickness of gate oxide (EOT) of MOSFET and power supply voltage (V_{dd}) will be also reduced to increase the transistor current I_{on} density (I_{on}/W) and decrease the circuit operation power. Smaller transistor and shorter internal connection will result in smaller capacitance, and these changes will all reduce the circuit delay time. Historically, the IC speed has grown by 30% at every new technology node. Higher speed will result in innovative IC applications, such as the higher CPU speed, higher DRAM and flash memory density, and higher broadband data transmission via RF circuit for cell phone. A good reviewing report is presented in Intel 22-nm FinFET study [2].

© Springer Nature Singapore Pte Ltd. 2018
Y.-C. Wu and Y.-R. Jhan, *3D TCAD Simulation for CMOS Nanoeletronic Devices*,
DOI 10.1007/978-981-10-3066-6_3

The most important advantages of scaling are as follows:

1. Reducing power consumption $P = kfCV_{dd}^2$, meaning that scaling can lead to reduced capacitance C and power supply voltage V_{dd}, thus effectively reducing power consumption;
2. The number of transistors in the chip will be doubled at every technology node, which can effectively reduce the cost;
3. Enhancing operating frequency. If there had been no scaling, doing the job of a single PC microprocessor chip (operating billions of transistors at 2 GHz) using old 1970 technology would require the power output of an electrical power generation plant. In short, for every technology, scaling improves cost, speed, and power consumption.

Technology innovation has made scaling feasible. Semiconductor researchers all over the world have been meeting several international conferences every year for generating consensus on transistor and circuit performance in order to meet future projected market demand in the future. International semiconductor organization updated annual report, **2015 International Technology Roadmap for Semiconductor (ITRS)** (Ref. [2] in Chap. 1) provides goals of scaling key performance index and pointed out the challenging issues of scaling.

The summary of 2015 ITRS is shown in Table 3.1. HP indicates high-performance technology, LSTP indicates low standby power technology for portable applications, and the actual gate length L_g reduces at every technology node. Take 16 nm as an example, even though the technology can fabricate the photoresist lines of 16 nm, engineers can convert the pattern into oxide lines. Then, use isotropic dry etching instrument to etch away oxide and to reduce the line width of oxide layer. Using this narrow oxide layer lines as the new etching mask, they can fabricate extremely small gate patterns by etching. Therefore, the IC scaling technology down to 5 nm node can be expectable based on dedicated innovation (Fig. 3.1).

In general, the scaling parameters of conventional MOSFET base on "**Constant Electric Field**" scaling principle which are as shown in Table 3.2, with the **scaling ratio constant $k = 0.7$**. It is shown in Table 3.2. In addition, for saving the area cost, the more important contributions of device scaling are improvement of device performance and reduction of circuit power consumption.

Along with the scaling of semiconductor device, various factors such as the severe short-channel effect (SCE), high leakage current (I_{off}), V_{th} roll-off, and V_{dd} cannot be reduced, and drain-induced barrier lowering (DIBL) has prevented 2D MOSFET from further scaling. Therefore, in recent years, all devices have been changed to 3D FinFET structure as shown in Fig. 3.2.

Table 3.1 Selected logic core device technology roadmap as predicted by 2015 ITRS version 2.0 (Ref. [2] in Chap. 1)

Year of production	2015	2017	2019	2021	2024	2027	2030
Logic device technology naming	P70M56	P48M36	P42M24	P32M20	P24M12G1	P24M12G2	P24M12G3
Logic industry "node range" labeling (nm)	"16/14"	"11/10"	"8/7"	"6/5"	"4/3"	"3/2.5"	"2/1.5"
Logic device structure options	FinFET FDSOI	FinFET FDSOI	FinFET LGAA	FinFET LGAA VGAA	VGAA,M3D	VGAA,M3D	VGAA,M3D
Logic device ground rules							
MPU/SoC Metalx V_2 pitch (nm)	28.0	18.0	12.0	10.0	6.0	6.0	6.0
MPU/SoC Metal0/1 V_i pitch (nm)	28.0	18.0	12.0	10.0	6.0	6.0	6.0
L_g: physical gate length for HP logic (nm)	24	18	14	10	10	10	10
L_g: physical gate length for LP logic (nm)	26	20	16	12	12	12	12
FinFET Fin width (nm)	8.0	6.0	6.0	NA	N/A	N/A	N/A
FinFET Fin height (nm)	42.0	42.0	42.0	NA	N/A	N/A	N/A
Device effective width (nm)	92.0	90.0	56.5	56.5	56.5	56.5	56.5
Device lateral half pitch (nm)	21.0	18.0	12.0	10.0	6.0	6.0	6.0
Device width or diameter (nm)	8.0	6.0	6.0	6.0	5.0	5.0	5.0
Device physical and electrical specs							
Power supply voltage V_{dd} (V)	0.80	0.75	0.70	0.65	0.55	0.45	0.40
Subthreshold slope (mV/dec)	75	70	68	65	40	25	25
Inversion layer thickness (nm)	1.10	1.00	0.90	0.85	0.80	0.80	0.80
$V_{t,sat}$ (mV) at I_{off} = 100 nA/μm (HP logic)	129	129	133	136	84	52	52
$V_{t,sat}$ (mV) at I_{off} = 100 pA/μm (LP logic)	351	336	333	326	201	125	125

(continued)

Table 3.1 (continued)

Year of production	2015	2017	2019	2021	2024	2027	2030
Effective mobility (cm^2/V s)	200	150	120	100	100	100	100
Rext (Ω μm)—HP logic	280	238	202	172	146	124	106
Ballistically injection velocity (cm/s)	1.20E−07	1.32E−07	1.45E−07	1.60E−07	1.76E−07	1.93E−07	2.13E−07
V_{dsat} (V)—HP logic	0.115	0.127	0.136	0.128	0.141	0.155	0.170
V_{dsat} (V)—LP logic	0.125	0.141	0.155	0.153	0.169	0.186	0.204
I_{on} (μA/μm) at I_{off} = 100nA/um—HP logic w/Rext = 0	2311	2541	2782	2917	3001	2670	2408

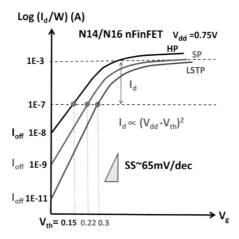

Fig. 3.1 I_d–V_g curve of N14/N16 HP, SP, LSTP nFinFET important parameters

Table 3.2 Device and circuits scaling key parameters

Device and circuit parameters		Scaling factor ($k < 1$, $k = 0.7$)
Scaled parameters	Device dimensions (L, t_{ox}, W, X_j)	k
	Doping concentration (N_a, N_d)	$1/k$
	Voltages	k
Effect on device parameters	Electric field	1
	Carrier velocity	1
	Depletion widths	k
	Capacitance ($C = \varepsilon A/t$)	k
	Drift current	k
Effect on circuit parameters	Device density	$1/k^2$
	Power density	1
	Power dissipation per device (P = VI)	k^2
	Circuit delay time ($\sim CV/I$)	k
	Power–delay product	k^3

3.2 Design Considerations of Threshold Voltage (V_{th}), Leakage Current (I_{off}), and Power Consumption (Power)

The circuit speed will be increased along with increasing I_{on}, thus requiring a smaller threshold voltage. However, the main current of MOSFET in the off state is I_{off}, and the I_d is the value of I_{off} measured with $V_{gs} = 0$ and $V_{ds} = V_{dd}$ as shown in

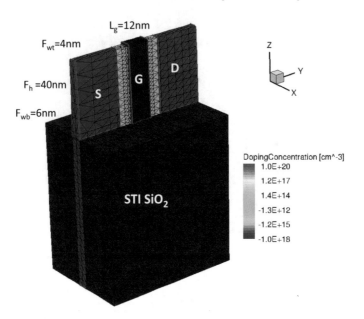

Fig. 3.2 L_g = 12 nm FinFET structure

Eq. (3.1). The minimization of I_{off} is the most important task for minimizing the power consumption by circuit in standby mode [1, 3].

Practically, **V_{th} is usually defined as the V_{gs} when I_{ds} = 100 nA $\times$ (W/L)** as shown in Eq. (3.1). MOSFET off state current (I_{off}) is defined by Eq. (3.2). The Eq. (3.1) can be substituted into obtain the equation of relationship between I_{off} and V_{th}, which is as shown as Eq. (3.3). Wherein, the simplification of Eq. (3.1) is another reason for V_{th} definition, meaning that the function $\exp(qV_{gs}/kT)$ will be changed by 10 whenever V_{gs} is changed by 60 mV under room temperature, such that $\exp(qV_{gs}/\eta kT)$ will be changed by 10 times at every $\eta \times 60$ mV. For example, if $\eta = 1.5$ and $V_{gs} < V_{th}$ under room temperature, it is indicated in Eq. (3.1) that I_{ds} will be reduced 10 times along with every decline of V_{gs} by 90 mV. $\eta \times 60$ mV is called subthreshold swing represented by the symbol SS as shown in Eq. (3.4) [1].

$$I_{ds}(\text{nA}) = 100 \cdot \frac{W}{L} e^{q(V_{gs}-Vt)/\eta kT} = 100 \cdot \frac{W}{L} \cdot 10^{(V_{gs}-Vt)/\text{SS}} \tag{3.1}$$

$$V_{ds} = 0, \quad I_{ds} = I_{off} \tag{3.2}$$

$$I_{off}(\text{nA}) = 100 \cdot \frac{W}{L} e^{-qVt/\eta kT} = 100 \cdot \frac{W}{L} \cdot 10^{-Vt/\text{SS}} \tag{3.3}$$

$$\text{SS(mV/decade)} = \eta \cdot 60 \, \text{mV} \tag{3.4}$$

The simplification of Eq. (3.2) is based on the exponential change of base as described below:

The exponential change of base is as follows:

$$e^a = 10^b, \ln e^a = \ln 10^b, \quad a = b \ln 10, \quad b = \frac{a}{\ln 10}$$

From (3.2)

$$a = \frac{q(V_{gs} - Vt)}{\eta \cdot kT} \therefore b = \frac{q(V_{gs} - V_t)}{\eta \cdot kT \cdot \ln 10} = \frac{(V_{gs} - V_t)}{\eta \cdot \left(\frac{kT}{q} \cdot \ln 10\right)} = \frac{(V_{gs} - V_t)}{SS} \tag{3.5}$$

$$SS = \eta \cdot \left(\frac{kT}{q} \cdot \ln 10\right) = \eta \cdot (26 \text{ mV} \times 2.3) = \eta \cdot 60 \text{ mV} \quad \text{at } 300 \text{ K} \tag{3.6}$$

$$SS = 60 \text{ mV} \cdot \eta \quad \because \eta = 1 + \frac{C_{dep}}{C_{oxe}} > 1 \quad \therefore \text{usually SS} > 60 \text{ mV} \tag{3.7}$$

As for the assigned W and L, there are two approaches for minimizing I_{off}. The first approach is to select higher V_{th}, yet this is not the optimal solution because higher V_{th} will lead to I_{on} reduction, thus lowering the circuit speed. Another better approach is to reduce subthreshold swing (SS) by increasing C_{oxe} to reduce η, which means the thinner gate oxide thickness (T_{ox}) is to be used. Using FinFET with high-k dielectric materials and metal gate can approach **ideal value SS = 60 mV/decade**. There is yet another approach for reducing I_{off} by SS reduction, which is allowing the transistor to be operated in low temperature. However, the low-temperature operation may lead to significant increase of cost.

V_{th} will be reduced along with the scaling of L as shown in Fig. 3.3. With significant reduction of V_{th}, I_{off} will become rather high, thus worsening the channel leakage current. For increasing V_{th}, the doping concentration of the body (N_b) of short-channel device will be higher than the long-channel device. The energy band diagram is shown for the long- and short-channel semiconductor–insulator junction in Fig. 3.3a, c with $V_{gs} = 0$. Figure 3.3b shows the case at $V_{gs} = V_{th}$. In the case of (b), E_c in the channel is pulled lower than in the case of (a), and therefore is closer to the E_c of source. In this case, electrons can flow from N^+ source through the channel to the drain. Figure 3.3c shows the case of short-channel device at $V_{gs} = 0$. If the channel is short enough, E_c will not be able to reach the same peak value as in (a). As the results, V_{th} value is lower in short-channel device than that of the long-channel device. The decreasing of V_{th} value can be explained as V_{th} roll-off. Therefore, V_{th} must be set in reasonable range for different gate lengths. Currently, for **FinFET N16/N14 node**, $\mathbf{V_{tn}}$ is **0.15–0.35 V** and $\mathbf{V_{tp}}$ is **−0.15 to −0.35 V** (Figs. 3.4 and 3.5).

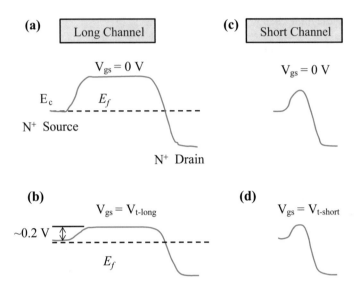

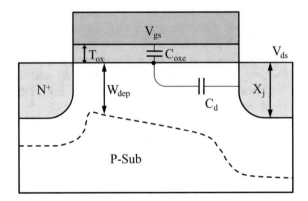

Fig. 3.3 a–d Energy band diagram of source to drain with V_{gs} = 0 V and V_{gs} = V_{th}, **a–b** long channel; **c–d** short channel

Fig. 3.4 Schematic two capacitor networks in MOSFET. C_d models the electrostatic coupling between the channel and the drain. As the channel length is reduced, drain to channel distance is reduced; therefore, C_d increases

3.3   Design Considerations of High-*k* Dielectric Materials and Metal Gate

With device size scaling, the thickness of gate oxide should be reduced along with the scaling of channel size. However, if the gate oxide is too thin, it will induce severe gate tunneling current, which will increase the device off current (I_{off}), thus leading to increased standby power consumption of the portable 3C products. Therefore, the severe gate leakage current will work against effective scaling down of device size.

Nowadays, the applications of high-*k* dielectric materials in semiconductor industry have been used [1]. With the requirements of reducing device dimension

Fig. 3.5 Drain could still have more control than the gate along another leakage current path below Si surface

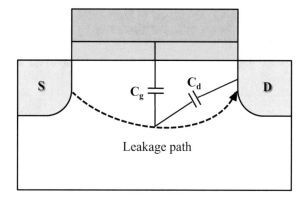

Fig. 3.6 Relationship between energy gaps (E_g) and relative dielectric constants (k) of various materials

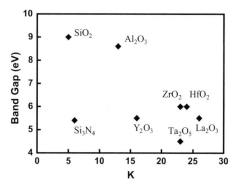

and gate leakage current, many gate dielectric materials with high-k value have been proposed to replace the traditional SiO$_2$ gate dielectric layer, including Al$_2$O$_3$, HfO$_2$, ZrO$_2$, and La$_2$O$_3$. Currently, in **N14/N16 FinFET,** the mainstream high-k dielectric material is **HfO$_2$ with relative dielectric constant k of 24.** The relationship between energy gaps (E_g) and relative dielectric constants (k) of various materials is as shown in Fig. 3.6.

With the same equivalent oxide thickness (EOT), the use of high-k material can reduce the gate leakage current, because the high-k material is higher than the traditional **SiO$_2$ dielectric constant of 3.9**, thus leading to higher relative capacitance in identical thickness. The higher gate capacitance increases higher driving current as shown in Eq. (3.8). The description and example are as shown below:

$$C_{\text{gate}} = \frac{k_{\text{ox}} \varepsilon_0}{t_{\text{ox}}} = \frac{k_{\text{hk}} \varepsilon_0}{t_{\text{hk}}} \ (\text{F/cm}^2) \tag{3.8}$$

$$\text{EOT} = \left(\frac{k_{\text{ox}}}{k_{\text{hk}}}\right) t_{\text{hk}} = t_{\text{ox}} \tag{3.9}$$

For example; For HfO$_2$, $t_{\text{hk}} = 6$ nm, $k_{\text{hk}} = 24$, $k_{\text{ox}} = 3.9$, $\therefore$ EOT $= 1$ nm

For N14/N16 FinFET, HfO$_2$: $t_{\text{hk}} = 3$ nm, then EOT $= 0.5$ nm

In addition to the impact of high-k material, the work functions (WF) of different metal gates being used will also affect V_{th} [4]; the relationship between metal work function and V_{th} can be substituted into V_{th} Eq. (3.10) via Eq. (3.11) in order to obtain the relationship equation as V_{tn} of nFinFET Eq. (3.12) and V_{tp} of pFinFET (3.13).

$$V_g = V_{fb} + \phi_s + V_{ox} \tag{3.10}$$

$$V_{fb} = V_{fb0} - Q_{ox} \Big/ C_{ox} = \psi_g - \psi_s - Q_{ox}/C_{ox} \tag{3.11}$$

$$V_{tn} = V_{fb} + \phi_S + V_{hk} = \psi_{mN} - \psi_S - \frac{Q_f}{C_{hk}} + \phi_S + V_{hk}$$
$$= \left(\psi_{mN} - \frac{Q_f}{C_{hk}} \right) - \psi_S + \phi_S + V_{hk} \tag{3.12}$$

$$V_{tp} = V_{fb} - \phi_S - V_{hk} = \psi_{mP} - \psi_S - \frac{Q_f}{C_{hk}} - \phi_S - V_{hk}$$
$$= \left(\psi_{mP} - \frac{Q_f}{C_{hk}} \right) - \psi_S - \phi_S - V_{hk} \tag{3.13}$$

ψ_{mN} is n-type metal WF;
ψ_{mP} is n-type metal WF;
ψ_S substrate WF;
Q_f is fixed charge in high-k material;
C_{hk} is capacitance of high-k material
ϕ_s is surface potential;
V_{hk} is voltage drop in high-k material.

The adjustable range of V_{th} of N14/N16 FinFET via channel doping has become rather small (<0.1 V). Thus, currently, different metal work functions are used for adjusting V_{tn} and V_{tp}. If the n-type metal is needed, the **low work function of Al-rich metal, like TiAl,** will be used, and if the p-type metal is needed, the **high work function of N-rich metal, like TiN,** will be used. The work function values of various metals frequently used for current semiconductor technologies are as shown in Fig. 3.7.

(1) **N-type Metal:**
   Al: (4.13 eV), Ta: (4.19 eV), Ti: (4.14 eV), Hf: (3.9 eV), TaN: (4.05 eV)
(2) **Midgap Metal:**
   W: (4.52 eV), Co: (4.45 eV), Pd: (4.9 eV), TiN: (4.7 eV), $TiSi_2$: (4.5 eV), TaN: (4.05 eV)
(3) **P-type Metal:**
   Pt: (5.65 eV), WN: (5.0 eV), Mo_2N: (5.33 eV), TaN: (5.43 eV), Ni: (5.2 eV)

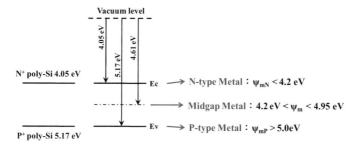

Fig. 3.7   Work function values of various metal frequently used in FinFET

3.4   Design Consideration of Device Gate and TCAD Design Guideline

The evolution of gate started from the initial planar gate to double gate and then advanced to current tri-gate and gate-all-around (GAA) as shown in Fig. 3.8. The more gate numbers covering the channel lead to better control capability [5]. The electrical line of electric field established by the multigate can be more focused and penetrating deeper in the channel to prevent electric field of drain. This effect can reduce the short-channel effect. The 3D electric field distribution within the transistor channel is as shown in Fig. 3.9.

From Poisson's equation, it is shown that the solution form of electrical potential of drain is as follows: $\varphi(x) = \varphi_0 \exp(-\frac{x}{\lambda_1})$, where λ **is defined as the natural length**.

A small λ value will result in rapidly reducing electrical potential of drain such that it will have lesser impact on the channel. This effect will help gate to dominate the transistor switching. The relationship equations of natural lengths under different gate structures are as shown in Table 3.3. The multigate structure like FinFET and gate-all-around FET (GAA FET) have smaller λ and leading to reduction of short-channel effect.

The smaller natural length will lead to better device performance. In general, the reasonable choice is $L_g > 5$–10 times of λ. We use the aforementioned equations derived from the square as the equivalent rectangular FinFET for approximation calculation. For example, in Fig. 3.10, $T_{si} \times T_{si} = F_w \times F_h$; the 10 nm × 10 nm square nanowire FET can be regarded as the equivalent of F_w of 7 nm × F_h of 14 nm or 5 nm × 20 nm rectangle FinFET, where λ_3 represents tri-gate (which is identical to FinFET) and λ_4 represents to GAA FET.

This chapter starts with the discussion of simulation of 3D FinFET, including the simulations of nFinFET, pFinFET I_d–V_g and I_d–V_d. First, the Predictive Technology Model for FinFETs is explained based on the paper published by ARM Company in ACM in 2012 [4], and the simulation process flow of its 3D FinFET is as shown in Fig. 3.11.

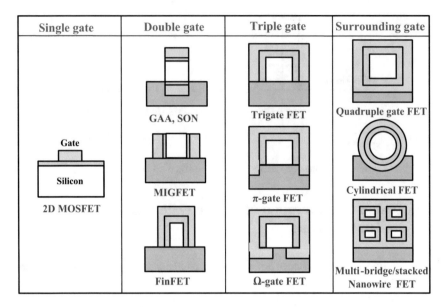

Single gate	Double gate	Triple gate	Surrounding gate
	GAA, SON	Trigate FET	Quadruple gate FET
Gate			
Silicon			
2D MOSFET	MIGFET	π-gate FET	Cylindrical FET
	FinFET	Ω-gate FET	Multi-bridge/stacked Nanowire FET

Fig. 3.8 Gate structure of FET evolution

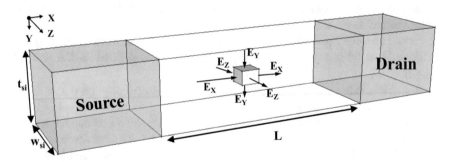

Fig. 3.9 Distribution of electric field within the channel of 3D transistor

Table 3.3 Natural lengths (λ) of different gate structures [5]

Single gate	$\lambda_1 = \sqrt{\dfrac{\varepsilon_{si}}{\varepsilon_{ox}} t_{si} t_{ox}}$
Double gate	$\lambda_2 = \sqrt{\dfrac{\varepsilon_{si}}{2\varepsilon_{ox}} t_{si} t_{ox}}$
Tri-gate (FinFET)	$\lambda_3 = \sqrt{\dfrac{\varepsilon_{si}}{3\varepsilon_{ox}} t_{si} t_{ox}}$
Quadruple gate	$\lambda_4 \cong \sqrt{\dfrac{\varepsilon_{si}}{4\varepsilon_{ox}} t_{si} t_{ox}}$
Surrounding gate (GAA)	$\lambda_0 \cong \sqrt{\dfrac{2\varepsilon_{si} t_{si}^2 \ln(1 + \frac{2t_{ox}}{t_{si}}) + \varepsilon_{ox} t_{si}^2}{16\varepsilon_{ox}}}$

Surrounding gate is also called gate-all-around (GAA)

Fig. 3.10 Natural length (λ) with the same T_{si} ($T_{si} = F_w = F_h$) in FinFET

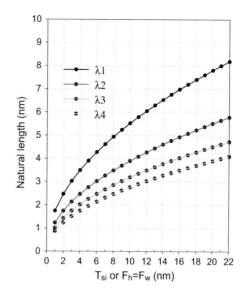

Fig. 3.11 Process flow of FinFET simulation. WF is metal gate work function

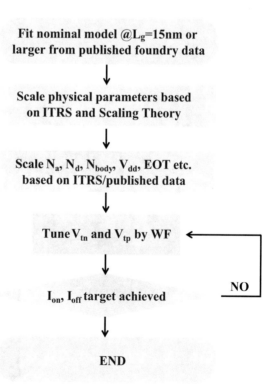

3.5   FinFET 3D Simulation

3.5.1   Establishment of FinFET Structure

The first thing is to start building the structure of 3D **FinFET** by right clicking commands on SDE tool to establish codes. The device length uses unit of nm. Before the establishment of 3D device structure, the **X–Y–Z coordinates and zero point** must be set firstly. Then, the "**eight cuboids**" form the basic structure of FinFET and describe in the following order:

(1) **Source contact (SC)**
(2) **Source (S)**
(3) **Gate oxide (Gox)**
(4) **Channel (channel)**
(5) **Drain (D)**
(6) **Drain contact (DC)**
(7) **Si Body (Body)**
(8) **STI buried oxide (Box)** (Fig. 3.12).

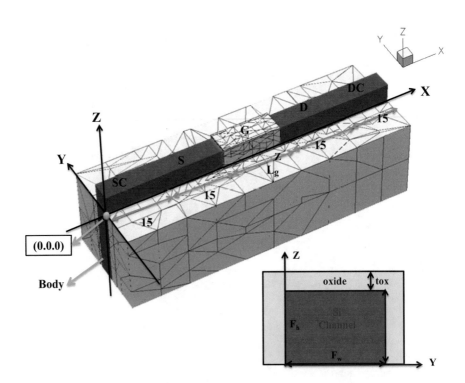

Fig. 3.12 Eight cuboids' structural diagram and 2D cross-sectional diagram of 3D FinFET structure from Synopsys Sentaurus screen capture. The unit of length is nm

Cuboids FinFET structures are created by sdegeo command of SDE tool. **Create-cuboid** refers to the size definition of 3D cuboid, which is determined by diagonal just like the establishment of 2D structure.

As shown in Fig. 3.14, the size of cuboid can be determined by assigning **vector from point A to point B**. For example, **(sdegeo:create-cuboid (position 0 0 0) (position x1 y1 z1) "Silicon" "SourceC") with coordinates of point A as (0, 0, 0) and coordinates of point B as (x1, y1, z1).** After the coordinates of point A and B are defined, the material of this cuboid should be created, and it is silicon in this example. After the material definition, the name of this cuboid should be defined, and in this example, it is defined as SourceC representing source contact. And then, the cuboids of Source, Channel, Drain, and Drain C should be established from left to right in accordance with Fig. 3.13 before the definition of all required contacts. Most examples in this chapter are not involved in the use of high-k material. Instead, the EOT equivalent is achieved by defining the ultra-thin $SiO_2 = 0.5$ nm. We do not set tunneling physical model along Z direction for simplification.

During the definition of 3D contact, only the equipotential surfaces such as electrode should be defined, and there is no need for defining the physical metal material and space of contact. In this case, once the coordinates of any given **any point onto the independently closed equipotential surface** is defined, the program will automatically extend leftward and rightward from that point until reaching the boundaries confining this closed contact surface region of 3D cuboid. For example, the code of **(sdegeo:set-contact-faces (find-face-id (position 1 1 z1)))** is for source contact (Fig. 3.13a **SourceC**), and **(sdegeo:set-contact-faces (find-face-id (position (+x2 1) 1 z2)))** is for top gate contact (Fig. 3.13a **Top Gate**).

3.5.2 Physical Property Analysis

After the FinFET cuboids are all established for simulation, the control gate (G) can be regarded as a conductor with voltage applied to the gate (V_g), and it is equipotential. Then, the metal gate work function (WK) must be defined in **devise_dvs.cmd** file, and **@WK@** must be added as a variable of SDEVICE tool of Sentaurus Workbench (SWB). The advantage of this approach is that multiple WK variables can be assigned for adjustment of V_{th} of 3D FinFET as shown in Fig. 3.15.

The physical and electrical properties can be examined during the analysis of FinFET device by Sentaurus. The **Sentaurus Visual tool** is as shown in Fig. 3.16, where the materials and their mesh can be selected to investigation via left toolbar. In addition, Sentaurus Visual tool allows the selection of physical property to be examined, such as energy band, electrons, and holes distributions. in toolbar. We take the energy band diagram for illustration. First, using the Y-axis cutting (Fig. 3.16), the 3D diagram converts to 2D diagram along Y-axis, and then the X–Z cross section as shown in Fig. 3.17.

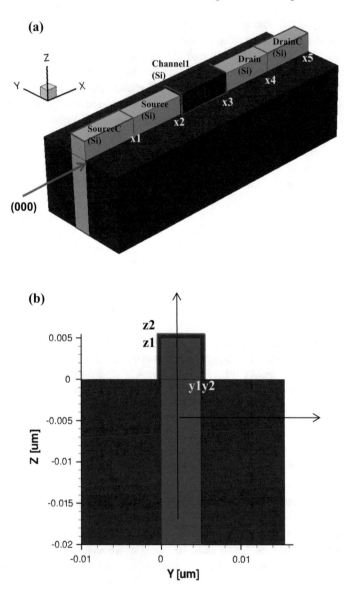

Fig. 3.13 Structure coordinated of **a** 3D FinFET structure and **b** Y–Z cross section of center of FinFET

And then, the variation of physical properties along X-axis should be inspected by cutting along Z-axis. The frame on the right side of the interface can be used to selected the 1D doping concentration in Fig. 3.18, and 1D energy band diagram in Fig. 3.19.

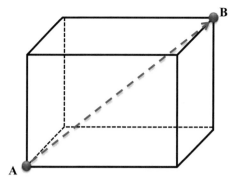

Fig. 3.14 Illustration of establishment of 3D cuboids (3D cuboid is formed by the diagonal from A to B)

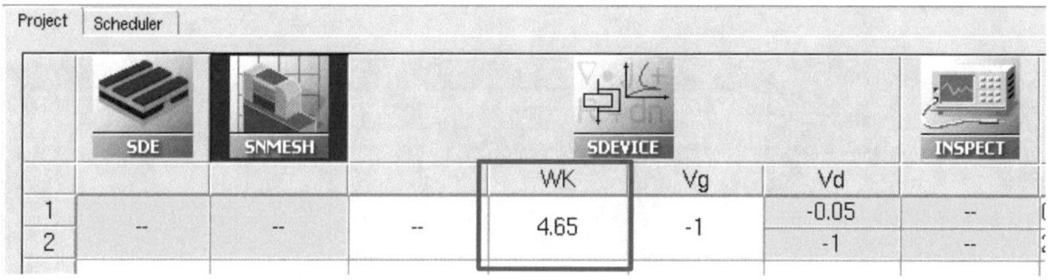

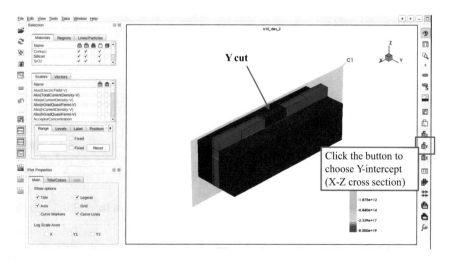

				WK	Vg	Vd		
1	--	--	--	4.65	-1	-0.05	--	
2						-1	--	

Fig. 3.15 Determination of variable values of @WK@ (4.65) metal work function, and gate voltage and drain voltage @V_g@ (−1), and @V_d@ (−0.05 and −1) within SWB

Fig. 3.16 3D structural diagram of n-type FinFET by Sentaurus Visual interface (readers can select the required FinFET physical property diagram from the toolbar on the *left*)

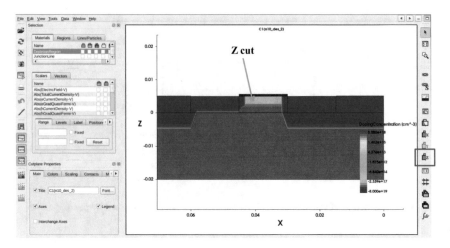

Fig. 3.17 2D structural diagram of n-type FinFET by Sentaurus Visual interface via toolbar

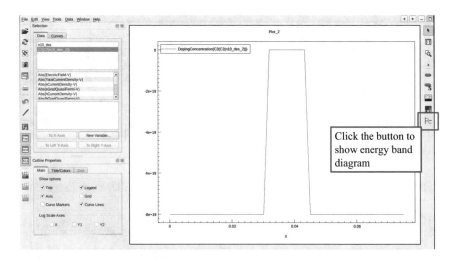

Fig. 3.18 1D structural diagram of doping concentration of n-type FinFET by Sentaurus Visual

Example 3.1 I_d–V_g **of 3D nFinFET with L_g = 15 nm**

The following three main program code files are based on Synopsys Sentaurus TCAD 2014 version. 1. SDE → devise_dvs.cmd, 2. SDEVICE → dessis_des.cmd, and 3. INSPECT → inspect_inc.cmd are discussed below (Figs. 3.20 and 3.21):

As well as Example 2.1 of Chap. 2, the devise_dvs.cmd can be divided into six parts.

(1) **Parameter**
(2) **Structure**

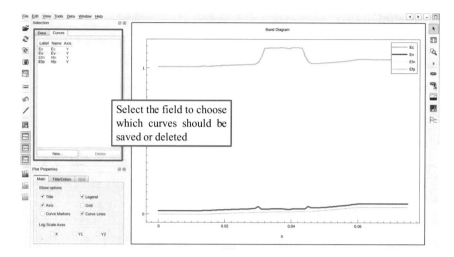

Fig. 3.19 Energy band diagram (E_c, E_v to X) of n-type FinFET by Sentaurus Visual interface

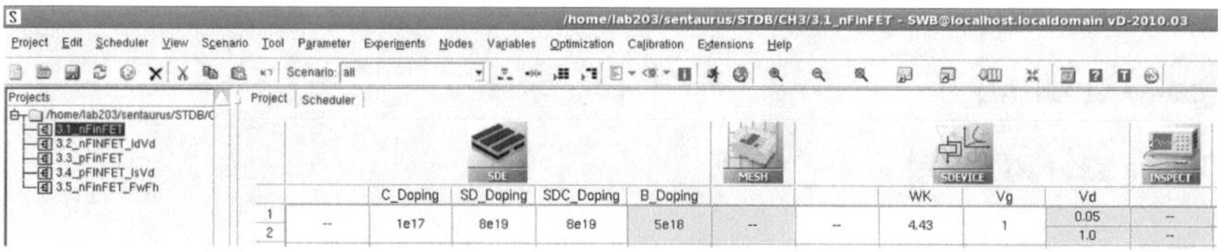

Fig. 3.20 Required simulation tools are shown in the workbench of 3D nFinFET TCAD simulation

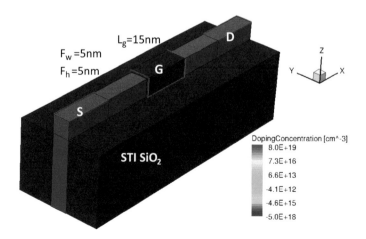

Fig. 3.21 Example 3.1, L_g = 15 nm 3D nFinFET TCAD simulation structure

(3) **Contact**
(4) **Doping**
(5) **Mesh**
(6) **Save**

1. SDE -- devise_dvs.cmd

;------------------ (1) parameter ----------------------;

(define nm 1e-3)

(define Fw 5)

(define Fh 5)

(define Lg 15)

(define LSDC 15)

(define LSD 15)

(define Tox 0.5)

(define x1 LSDC)

(define x2 (+ x1 LSD))

(define x3 (+ x2 Lg))

(define x4 (+ x3 LSD))

(define x5 (+ x4 LSDC))

(define y1 Fw)

(define y2 (+ y1 Tox))

(define y3 (+ y2 10))

(define z1 Fh)

(define z2 (+ z1 Tox))

(define C_Doping @C_Doping@)

;(define C_Doping 1e11)

(define SD_Doping @SD_Doping@)

(define SDC_Doping @SDC_Doping@)

;(define B_Doping 1e15)

(define B_Doping @B_Doping@)

```
; ---------------------- (2) Structure ---------------------;

"ABA"

;--- Source contact and Source ---;

(sdegeo:create-cuboid (position 0 0 0 ) (position x1 y1 z1 ) "Silicon" "SourceC")

(sdegeo:create-cuboid (position x1 0 0 ) (position x2 y1 z1) "Silicon" "Source")

;--- Gate oxide ---;

(sdegeo:create-cuboid (position x2 (- Tox) 0 ) (position  x3 y2 z2 ) "SiO2" "Gateoxide")

;--- Channel ---;

(sdegeo:create-cuboid (position x2 0 0 ) (position x3 y1 z1 ) "Silicon" "Channel")

;--- Drain contact and Drain---;

(sdegeo:create-cuboid (position x3 0 0 ) (position x4 y1 z1 ) "Silicon" "Drain")

(sdegeo:create-cuboid (position x4 0 0 ) (position x5 y1 z1 ) "Silicon" "DrainC")

;--- Buried oxide ---;

(sdegeo:create-cuboid (position 0 (- 10) (- 20) ) (position x5 y3 0 ) "SiO2" "Box")

"ABA"

;--- Si Body ---;

(sdegeo:create-cuboid (position 0 0 (- 20) ) (position x5 y1 0 ) "Silicon" "Body")

; -------------------- (3) Contact ------------------;

;----- Source -----;

(sdegeo:define-contact-set "S" 4.0  (color:rgb 1.0 0.0 0.0 ) "##" )

(sdegeo:set-current-contact-set "S")

(sdegeo:set-contact-faces (find-face-id (position 1 1 z1)))

;----- Drain -----;

(sdegeo:define-contact-set "D" 4.0  (color:rgb 1.0 0.0 0.0 ) "##" )

(sdegeo:set-current-contact-set "D")
```

(sdegeo:set-contact-faces (find-face-id (position (+ x4 1) 1 z1)))

;----- Front Gate -----;

(sdegeo:define-contact-set "G" 4.0 (color:rgb 1.0 0.0 0.0) "||")

(sdegeo:set-current-contact-set "G")

(sdegeo:set-contact-faces (find-face-id (position (+ x2 1) (- Tox) 1)))

;----- Top Gate -----;

(sdegeo:define-contact-set "G" 4.0 (color:rgb 1.0 0.0 0.0) "||")

(sdegeo:set-current-contact-set "G")

(sdegeo:set-contact-faces (find-face-id (position (+ x2 1) 1 z2)))

;----- Back Gate -----;

(sdegeo:define-contact-set "G" 4.0 (color:rgb 1.0 0.0 0.0) "||")

(sdegeo:set-current-contact-set "G")

(sdegeo:set-contact-faces (find-face-id (position (+ x2 1) y2 1)))

;----- Body -----;

(sdegeo:define-contact-set "B" 4.0 (color:rgb 1.0 0.0 0.0) "##")

(sdegeo:set-current-contact-set "B")

(sdegeo:set-contact-faces (find-face-id (position (* 0.5 x5) (* 0.5 y1) (- 20))))

; ----------------------- (4) Doping ------------------------;

;----- Channel -----;

(sdedr:define-constant-profile "dopedC" "BoronActiveConcentration" C_Doping)

(sdedr:define-constant-profile-region "RegionC" "dopedC" "Channel")

;----- Source -----;

(sdedr:define-constant-profile "dopedS" "ArsenicActiveConcentration" SD_Doping)

(sdedr:define-constant-profile-region "RegionS" "dopedS" "Source")

(sdedr:define-constant-profile "dopedSC" "ArsenicActiveConcentration" SDC_Doping)

(sdedr:define-constant-profile-region "RegionSC" "dopedSC" "SourceC")

;----- Drain ------;

(sdedr:define-constant-profile "dopedD" "ArsenicActiveConcentration" SD_Doping)

(sdedr:define-constant-profile-region "RegionD" "dopedD" "Drain")

(sdedr:define-constant-profile "dopedDC" "ArsenicActiveConcentration" SDC_Doping)

(sdedr:define-constant-profile-region "RegionDC" "dopedDC" "DrainC")

;----- Si Body -----;

(sdedr:define-constant-profile "dopedB" "BoronActiveConcentration" B_Doping)

(sdedr:define-constant-profile-region "RegionB" "dopedB" "Body")

; -------------------- (5) Mesh ----------------------;

;--- AllMesh ---;

(sdedr:define-refinement-size "Cha_Mesh" 5 5 5 1 1 1)

(sdedr:define-refinement-material "channel_RF" "Cha_Mesh" "Silicon")

;--- ChannelMesh ---;

(sdedr:define-refinement-window "multiboxChannel" "Cuboid"

(position x1 0 0)   (position x4 y1 z1))

(sdedr:define-multibox-size "multiboxSizeChannel" 2 2 2 2 2 2)

(sdedr:define-multibox-placement "multiboxPlacementChannel" "multiboxSizeChannel" "multiboxChannel")

(sdedr:define-refinement-function "multiboxPlacementChannel" "DopingConcentration" "MaxTransDiff" 1)

; ----------- (6) Save (BND and CMD and rescale to nm) -----------;

(sde:assign-material-and-region-names (get-body-list))

(sdeio:save-tdr-bnd (get-body-list) "n@node@_nm.tdr")

(sdedr:write-scaled-cmd-file "n@node@_msh.cmd" nm)

(define sde:scale-tdr-bnd

 (lambda (tdrin sf tdrout)

  (sde:clear)

  (sdegeo:set-default-boolean "XX")

  (sdeio:read-tdr-bnd tdrin)

  (entity:scale (get-body-list) sf)

  (sdeio:save-tdr-bnd (get-body-list) tdrout)

  ))

(sde:scale-tdr-bnd "n@node@_nm.tdr" nm "n@node@_bnd.tdr")

; -------------------------------------- END --------------------------------------;

The "ABA" is important command to define gate insulator and channel region for FinFET or other complex device structures. It defines the latter (or new) cuboid replacing former (or old) cuboide in their overlapping region. Figure 3.22 illustrates the "ABA" command results. On the other hand, "BAB" command can use for former (or old) cuboide replacing latter (or new) cuboide in their overlapping region.

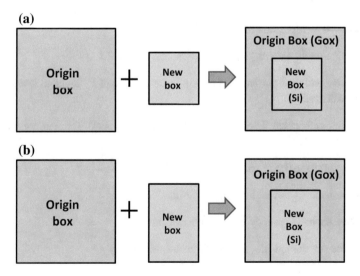

Fig. 3.22 Illustration of "ABA" command. It is very useful in 3D FET, **a** is suitable for gate-all-around FET, and **b** FinFET

2. SDEVICE -- dessis_des.cmd

```
#-------------------------- dessis_des.cmd --------------------------------#
File{

    Grid="@tdr@"

    Plot="@tdrdat@"

    Current="@plot@"

    Output="@log@"

}

  Electrode {

   { name="S"    Voltage=0.0 }

   { name="D"    Voltage=0.0 }

   { name="G"     Voltage=0 WorkFunction=@WK@}

   { name="B"    Voltage=0.0 }

  }

Physics{

   Mobility( DopingDep HighFieldSaturation Enormal )

   EffectiveIntrinsicDensity( OldSlotboom )

   Recombination( SRH(DopingDep) )

   }

Math{

  Extrapolate

  Derivatives

  * Avalderivatives

  RelErrControl

  Digits=5
```

```
   ErRef(electron)=1.e10

   ErRef(hole)=1.e10

   Notdamped=50

   Iterations=20

   *Newdiscretization
   Directcurrent

   Method=ParDiSo

   Parallel= 2

   *-VoronoiFaceBoxMethod

   NaturalBoxMethod

}
Plot{

  eDensity hDensity

  eCurrent hCurrent

  TotalCurrent/Vector eCurrent/Vector hCurrent/Vector

  eMobility hMobility

  eVelocity hVelocity

  eEnormal hEnormal

  ElectricField/Vector Potential SpaceCharge

  eQuasiFermi hQuasiFermi

  Potential Doping SpaceCharge

  SRH Auger

  AvalancheGeneration

  DonorConcentration AcceptorConcentration

  Doping

  eGradQuasiFermi/Vector hGradQuasiFermi/Vector
```

    eEparallel hEparalllel

    BandGap

    BandGapNarrowing

    Affinity

    ConductionBand ValenceBand

    }

Solve{

NewCurrentFile=""

Coupled(Iterations=100){ Poisson }

Coupled(Iterations=100){ Poisson Electron Hole }

Coupled{ Poisson Electron Hole }

Quasistationary(

InitialStep=0.01 Increment=1.35

MinStep=1e-5 MaxStep=0.2

Goal{ Name="D" Voltage= @Vd@ }

){ Coupled{ Poisson Electron Hole } }

Quasistationary(

InitialStep=1e-3 Increment=1.35

MinStep=1e-5 MaxStep=0.05

Goal{ Name="G" Voltage= @Vg@ }

){ Coupled{ Poisson Electron Hole} }

}

#---------------------------- END --------------------------------#

3. INSPECT -- inspect_inc.cmd

```
#------------------------------------------------------------------------#

#       Script file designed to compute  :        #

#        * The threshold voltage          : VT   #

#        * The transconductance           : gm   #

#------------------------------------------------------------------------ #

if { ! [catch {open n@previous@_ins.log w} log_file] } {

        set fileId stdout

}

puts $log_file " "

puts $log_file "            ------------------------------------ "

puts $log_file "            Values of the extracted Parameters : "

puts $log_file "            ------------------------------------ "

puts $log_file " "

puts $log_file " "

set  DATE  [ exec  date ]

set  WORK  [ exec pwd   ]

puts $log_file " Date     : $DATE "

puts $log_file " Directory : $WORK "

puts $log_file " "

puts $log_file " "

#     idvgs=y(x) ;  vgsvgs=x(x) ;   #

set out_file n@previous@_des
```

```
proj_load "${out_file}.plt"

# ---------------------------------------------------------------------- #

# I) VT = Xintercept(maxslope(ID[VGS])) or VT = VGS( IDS= 0.1 ua/um ) #

# ---------------------------------------------------------------------- #

cv_create   idvgs  "${out_file} G OuterVoltage" "${out_file} D TotalCurrent"

cv_create   vdsvgs "${out_file} G OuterVoltage" "${out_file} D OuterVoltage"

#.................................................. #

# 1) VT extracted as the intersection point with the X axis at the point #

#    where the id(vgs) slope reaches its maxmimum :        #

#.................................................. #

set VT1   [ f_VT1 idvgs ]

#..........................................        #

# 2) Printing of the whole set of extracted values (std output) : #

#..........................................        #

puts $log_file "Threshold   voltage VT1 = $VT1 Volts"

puts $log_file " "

#.................................................. #

# 3) Initialization and display of curves on the main Inspect screen  : #

# .................................................. #

cv_display    idvgs

cv_lineStyle idvgs  solid

cv_lineColor idvgs  red

# ---------------------------------------------------------------------- #
```

II)   gm = maxslope((ID[VGS])

--

set gm   [f_gm idvgs]

puts $log_file " "

puts $log_file "Transconductance gm   = $gm A/V"

puts $log_file " "

set ioff [cv_compute "vecmin(<idvgs>)" A A A A]

puts $log_file " "

puts $log_file "Current ioff   = $ioff A"

puts $log_file " "

set isat [cv_compute "vecmax(<idvgs>)" A A A A]

puts $log_file " "

puts $log_file "Current isat   = $isat A"

puts $log_file " "

set rout [cv_compute "Rout(<idvgs>)" A A A A]

puts $log_file " "

puts $log_file "Resistant rout   = $rout A"

puts $log_file " "

cv_createWithFormula logcurve "log10(<idvgs>)" A A A A

cv_createWithFormula difflog "diff(<logcurve>)" A A A A

set sslop [cv_compute "1/vecmax(<difflog>)" A A A A]
puts $log_file " "

puts $log_file "sub solp   = $sslop A/V"

puts $log_file " "

Puting into Family Table

ft_scalar VT $VT1

ft_scalar gmax $gm

ft_scalar ioff $ioff

ft_scalar isat $isat

ft_scalar sslop $sslop

ft_scalar rout $rout

close $log_file

#------------------------------------- END --#

The electric property diagram of I_d–V_g of nFinFET simulation result is as shown in Fig. 3.23. The important parameters are SS at around 67 mV/dec, V_{th} at around −0.27 V, I_{sat} at around 2.43 × 10^{-5} A, and I_{off} as around 8.35 × 10^{-12} A as shown in Fig. 3.24. The structural channel mesh, the electron concentration distributions of 3D and 2D structures, electric field distributions, electric potential distributions, and the energy band diagrams along the channel direction are as shown in Figs. 3.25, 3.26, 3.27, 3.28, 3.29, 3.30, 3.31 and 3.32 with the conditions of L_g = 15 nm, V_d = 1 V, and V_g = 1 V. The Sentaurus TCAD simulation result is very close to the Intel 14 nm experimental result.

Example 3.2 I_d–V_g of 3D nFinFET
The following three main program code files are based on Synopsys Sentaurus TCAD 2014 version (Fig. 3.33).

3D -nFinFET

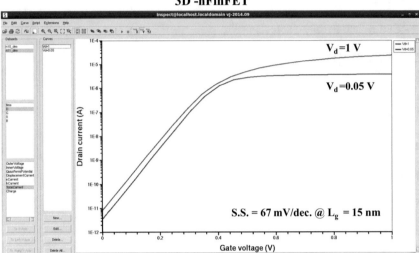

Fig. 3.23 I_d–V_g curve of 3D nFinFET simulation, some of the descriptions are added by PowerPoint after snapshot by inspect tool

	VT	gmax	Ioff	isat	sslop	rout
1	--	0.2991801346321319	026380288151632e-	3.60431201992297e-12	3.83627999292586e-06	0.06673700664293655	8355.687665-
2	--	0.2749392293301869	163211637402651e-	8.34712177542617e-12	2.42589227332663e-05	0.06787417727516247	11382.606870:

Fig. 3.24 Electric property parameters of 3D nFinFET simulation

3D –nFinFET Mesh

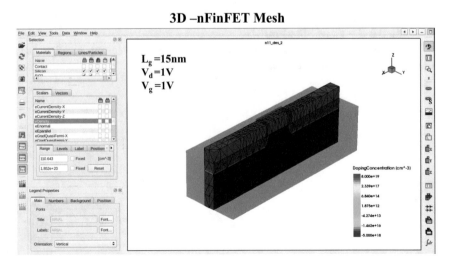

Fig. 3.25 Mesh diagram of 3D nFinFET simulation

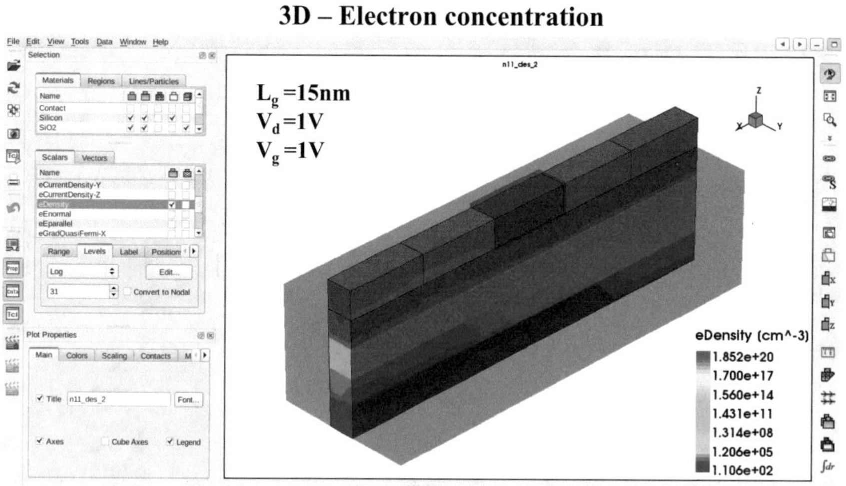

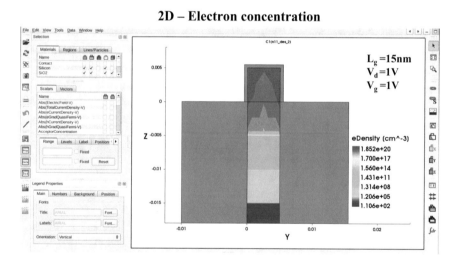

Fig. 3.26 Electron concentration distribution of 3D nFinFET simulation

Fig. 3.27 Electron concentration distribution of 2D nFinFET simulation

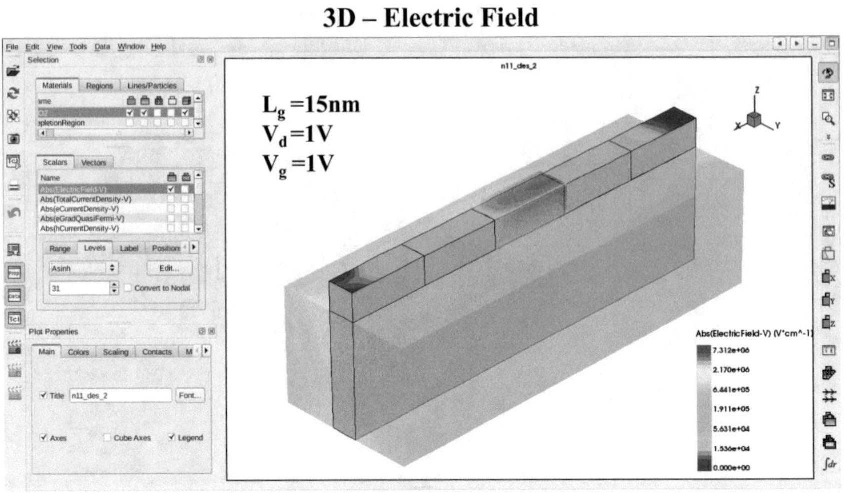

3D – Electric Field

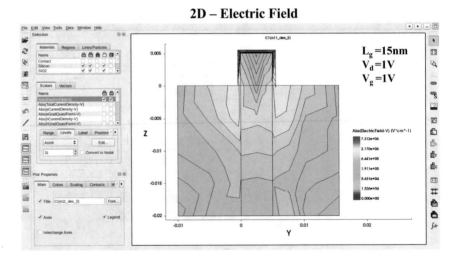

Fig. 3.28 Electric field distribution of 3D nFinFET simulation

2D – Electric Field

Fig. 3.29 Electric field distribution of 2D nFinFET simulation

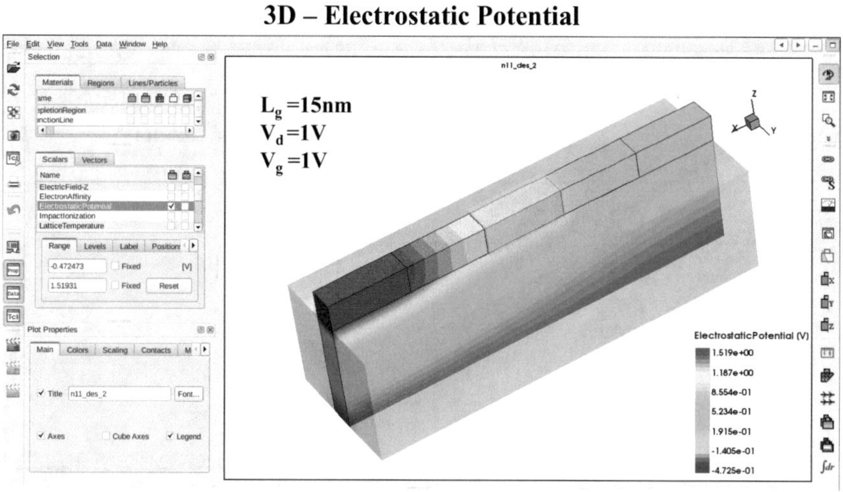

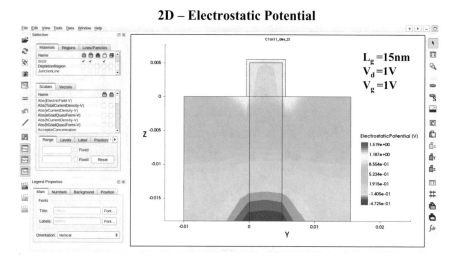

Fig. 3.30 Electric potential distribution of 3D nFinFET simulation

Fig. 3.31 Electric potential distribution of 2D nFinFET simulation

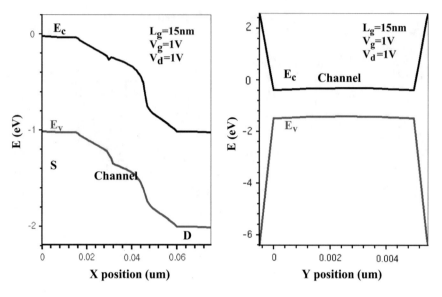

Fig. 3.32 Energy band diagram E_c and E_v to **a** X and **b** Y of 3D nFinFET simulation

Fig. 3.33 Required simulation tools are shown in the workbench of nFinFET simulation

The SDE – devise_dvs.cmd, SDEVICE – dessis_des.cmd, and INSPECT – inspect_inc.cmd are as shown below:

1. SDE -- devise_dvs.cmd

```
;----- parameter -----;

(define nm 1e-3)

(define Fw 5)

(define Fh 5)

(define Lg 15)

(define LSDC 15)

(define LSD 15)

(define Tox 0.5)

(define x1 LSDC)

(define x2 (+ x1 LSD))

(define x3 (+ x2 Lg))

(define x4 (+ x3 LSD))
(define x5 (+ x4 LSDC))

(define y1 Fw)

(define y2 (+ y1 Tox))

(define y3 (+ y2 10))

(define z1 Fh)

(define z2 (+ z1 Tox))

(define C_Doping @C_Doping@)

(define SD_Doping @SD_Doping@)

(define SDC_Doping @SDC_Doping@)

;(define B_Doping 1e15)

(define B_Doping @B_Doping@)

; ----- Structure -----;
```

"ABA"

;--- Source contact and Source ---;

(sdegeo:create-cuboid (position 0 0 0) (position x1 y1 z1) "Silicon" "SourceC")

(sdegeo:create-cuboid (position x1 0 0) (position x2 y1 z1) "Silicon" "Source")

;--- Gate oxide ---;

(sdegeo:create-cuboid (position x2 (- Tox) 0) (position x3 y2 z2) "SiO2" "Gateoxide")

;--- Channel ---;

(sdegeo:create-cuboid (position x2 0 0) (position x3 y1 z1) "Silicon" "Channel")

;--- Drain contact and Drain---;

(sdegeo:create-cuboid (position x3 0 0) (position x4 y1 z1) "Silicon" "Drain")

(sdegeo:create-cuboid (position x4 0 0) (position x5 y1 z1) "Silicon" "DrainC")

;--- Buried oxide ---;

(sdegeo:create-cuboid (position 0 (- 10) (- 20)) (position x5 y3 0) "SiO2" "Box")

"ABA"

;--- Si Body ---;

(sdegeo:create-cuboid (position 0 0 (- 20)) (position x5 y1 0) "Silicon" "Body")

; ----- Contact-----;

;----- Source -----;

(sdegeo:define-contact-set "S" 4.0 (color:rgb 1.0 0.0 0.0) "##")

(sdegeo:set-current-contact-set "S")

(sdegeo:set-contact-faces (find-face-id (position 1 1 z1)))

;----- Drain -----;

(sdegeo:define-contact-set "D" 4.0 (color:rgb 1.0 0.0 0.0) "##")

(sdegeo:set-current-contact-set "D")

(sdegeo:set-contact-faces (find-face-id (position (+ x4 1) 1 z1)))

```
;----- Front Gate -----;

(sdegeo:define-contact-set "G" 4.0  (color:rgb 1.0 0.0 0.0 ) "||" )

(sdegeo:set-current-contact-set "G")

(sdegeo:set-contact-faces (find-face-id (position (+ x2 1) (- Tox) 1 )))

;----- Top Gate -----;

(sdegeo:define-contact-set "G" 4.0  (color:rgb 1.0 0.0 0.0 ) "||" )

(sdegeo:set-current-contact-set "G")

(sdegeo:set-contact-faces (find-face-id (position (+ x2 1) 1  z2 )))

;----- Back Gate -----;

(sdegeo:define-contact-set "G" 4.0  (color:rgb 1.0 0.0 0.0 ) "||" )

(sdegeo:set-current-contact-set "G")

(sdegeo:set-contact-faces (find-face-id (position (+ x2 1) y2 1 )))

;----- Body -----;

(sdegeo:define-contact-set "B" 4.0  (color:rgb 1.0 0.0 0.0 ) "##" )

(sdegeo:set-current-contact-set "B")

(sdegeo:set-contact-faces (find-face-id (position (* 0.5 x5) (* 0.5 y1) (- 20) )))

; ----- Doping-----;

;----- Channel -----;

(sdedr:define-constant-profile "dopedC" "BoronActiveConcentration" C_Doping )
(sdedr:define-constant-profile-region  "RegionC" "dopedC" "Channel" )

;----- Source -----;

(sdedr:define-constant-profile "dopedS" "ArsenicActiveConcentration" SD_Doping )

(sdedr:define-constant-profile-region  "RegionS" "dopedS" "Source" )

(sdedr:define-constant-profile "dopedSC" "ArsenicActiveConcentration" SDC_Doping )

(sdedr:define-constant-profile-region  "RegionSC" "dopedSC" "SourceC" )
```

;----- Drain ------;

(sdedr:define-constant-profile "dopedD" "ArsenicActiveConcentration" SD_Doping)

(sdedr:define-constant-profile-region "RegionD" "dopedD" "Drain")

(sdedr:define-constant-profile "dopedDC" "ArsenicActiveConcentration" SDC_Doping)

(sdedr:define-constant-profile-region "RegionDC" "dopedDC" "DrainC")

;----- Si Body -----;

(sdedr:define-constant-profile "dopedB" "BoronActiveConcentration" B_Doping)

(sdedr:define-constant-profile-region "RegionB" "dopedB" "Body")

; ----- Mesh -----;

;--- AllMesh ---;

(sdedr:define-refinement-size "Cha_Mesh" 5 5 5 1 1 1)

(sdedr:define-refinement-material "channel_RF" "Cha_Mesh" "Silicon")

;--- ChannelMesh ---;

(sdedr:define-refinement-window "multiboxChannel" "Cuboid"

(position x1 0 0)   (position x4 y1 z1))

(sdedr:define-multibox-size "multiboxSizeChannel" 2 2 2 2 2 2)

(sdedr:define-multibox-placement "multiboxPlacementChannel" "multiboxSizeChannel"
"multiboxChannel")

(sdedr:define-refinement-function "multiboxPlacementChannel" "DopingConcentration"
"MaxTransDiff" 1)

; ----- Save BND and CMD and rescale to nm -----;

(sde:assign-material-and-region-names (get-body-list))

(sdeio:save-tdr-bnd (get-body-list) "n@node@_nm.tdr")
(sdedr:write-scaled-cmd-file "n@node@_msh.cmd" nm)

(define sde:scale-tdr-bnd

 (lambda (tdrin sf tdrout)

 (sde:clear)

 (sdegeo:set-default-boolean "XX")

 (sdeio:read-tdr-bnd tdrin)

 (entity:scale (get-body-list) sf)

 (sdeio:save-tdr-bnd (get-body-list) tdrout)

)

)

(sde:scale-tdr-bnd "n@node@_nm.tdr" nm "n@node@_bnd.tdr")

; ----- END -----;

2. SDEVICE -- dessis_des.cmd

File{

    Grid="@tdr@"

    Plot="@tdrdat@"

    Current="@plot@"

    Output="@log@"

}

Electrode {

   { name="S"   Voltage=0.0 }

   { name="D"   Voltage=0.0 }

   { name="G"   Voltage=0 WorkFunction=@WK@}

   { name="B"   Voltage=0.0 }

  }

Physics{

   Mobility(DopingDep HighFieldSaturation Enormal)

   EffectiveIntrinsicDensity(OldSlotboom)

   Recombination(SRH(DopingDep))

   }

 Math{

   Extrapolate

   Derivatives

   * Avalderivatives

   RelErrControl

   Digits=5

   ErRef(electron)=1.e10

   ErRef(hole)=1.e10

   Notdamped=50

   Iterations=20

```
  *Newdiscretization

  Directcurrent

  Method=ParDiSo

  Parallel= 2

  *-VoronoiFaceBoxMethod

  NaturalBoxMethod

}

Plot{

 eDensity hDensity

 eCurrent hCurrent

 TotalCurrent/Vector eCurrent/Vector hCurrent/Vector

 eMobility hMobility

 eVelocity hVelocity

 eEnormal hEnormal

  ElectricField/Vector Potential SpaceCharge

  eQuasiFermi hQuasiFermi

  Potential Doping SpaceCharge

  SRH Auger

  AvalancheGeneration

  DonorConcentration AcceptorConcentration

  Doping
```

eGradQuasiFermi/Vector hGradQuasiFermi/Vector

eEparallel hEparalllel

BandGap

BandGapNarrowing

Affinity

ConductionBand ValenceBand

}

Solve{

NewCurrentFile=""

Coupled(Iterations=100){ Poisson }

Coupled(Iterations=100){ Poisson Electron Hole }

Coupled{ Poisson Electron Hole }

Quasistationary(

InitialStep=0.01 Increment=1.35

MinStep=1e-5 MaxStep=0.2

Goal{ Name="G" Voltage= @Vg@ }

){ Coupled{ Poisson Electron Hole} }

Quasistationary(

InitialStep−1e-3 Increment−1.35

MinStep=1e-5 MaxStep=0.05

Goal{ Name="D" Voltage= @Vd@ }

 ){ Coupled{ Poisson Electron Hole } }

}

----- END -----

3. INSPECT -- inspect_inc.cmd

```
#----------------------------------------------------------------------#
#      Script file designed to compute   :         #
#        * The threshold voltage          : VT   #
#        * The transconductance           : gm  #
#----------------------------------------------------------------------#


if { ! [catch {open n@previous@_ins.log w} log_file] } {

          set fileId stdout

}
puts $log_file " "
puts $log_file "             ----------------------------------- "
puts $log_file "             Values of the extracted Parameters : "
puts $log_file "             ----------------------------------- "
puts $log_file " "
puts $log_file " "
set  DATE  [ exec  date ]
set  WORK  [ exec pwd  ]
puts $log_file " Date     : $DATE "
puts $log_file " Directory : $WORK "
puts $log_file " "
puts $log_file " "
#              idvgs=y(x) ;  vgsvgs=x(x) ;  #
```

```
set out_file n@previous@_des

proj_load "${out_file}.plt"

# ----------------------------------------------------------------------- #

# I)  VT = Xintercept(maxslope(ID[VGS]))  or  VT = VGS( IDS= 0.1 ua/um ) #

# ----------------------------------------------------------------------- #

cv_create   idvgs  "${out_file} G OuterVoltage" "${out_file} D TotalCurrent"

cv_create   vdsvgs "${out_file} G OuterVoltage" "${out_file} D OuterVoltage"

#.................................................... #

# 1) VT extracted as the intersection point with the X axis at the point  #

#    where the id(vgs) slope reaches its maxmimum :            #

#.................................................... #

set VT1   [ f_VT1 idvgs ]

#........................................... #

# 2) Printing of the whole set of extracted values (std output) :  #

#........................................... #

puts $log_file "Threshold   voltage VT1 = $VT1 Volts"

puts $log_file " "

#.................................................... #

# 3) Initialization and display of curves on the main Inspect screen  : #

# .................................................. #

cv_display    idvgs

cv_lineStyle idvgs  solid

cv_lineColor idvgs  red
```

```
# --------------------------------------------------------------------- #
# II)                    gm =  maxslope((ID[VGS])        #
# --------------------------------------------------------------------- #
set gm    [ f_gm idvgs ]
puts $log_file " "
puts $log_file "Transconductance gm                          = $gm  A/V"
puts $log_file " "
set ioff  [ cv_compute "vecmin(<idvgs>)" A A A A ]
puts $log_file " "
puts $log_file "Current ioff       = $ioff  A"
puts $log_file " "
set isat  [ cv_compute "vecmax(<idvgs>)" A A A A ]
puts $log_file " "
puts $log_file "Current isat       = $isat  A"
puts $log_file " "
set rout  [ cv_compute "Rout(<idvgs>)" A A A A ]
puts $log_file " "
puts $log_file "Resistant rout       = $rout  A"
puts $log_file " "
cv_createWithFormula logcurve "log10(<idvgs>)" A A A A
cv_createWithFormula difflog "diff(<logcurve>)" A A A A
set sslop [ cv_compute "1/vecmax(<difflog>)" A A A A ]
puts $log_file " "
```

```
puts $log_file "sub solp        = $sslop  A/V"

puts $log_file " "

ft_scalar VT $VT1

ft_scalar gmax $gm

ft_scalar ioff $ioff

ft_scalar isat $isat

ft_scalar sslop $sslop

ft_scalar rout $rout

close $log_file

# ----- END -----#
```

The I_d–V_d diagram of 3D nFinFET output curves simulation result is as shown in Fig. 3.34. With V_{tn} =+0.3 V, when V_g = 0.2 V which is less than V_{tn} = 0.3 V, the output curves of V_g = 0.2 V are zero.

Example 3.3 I_d–V_g of 3D pFinFET

The following three main program code files are based on Synopsys Sentaurus TCAD 2014 version. 1. SDE → devise_dvs.cmd, 2. SDEVICE → dessis_des.cmd, and 3. INSPECT → inspect_inc.cmd are discussed below (Fig. 3.35):

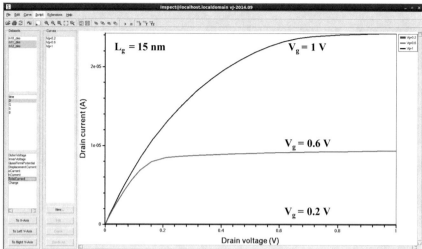

Fig. 3.34 I_d–V_d curve of 3D nFinFET simulation with L_g = 15 nm

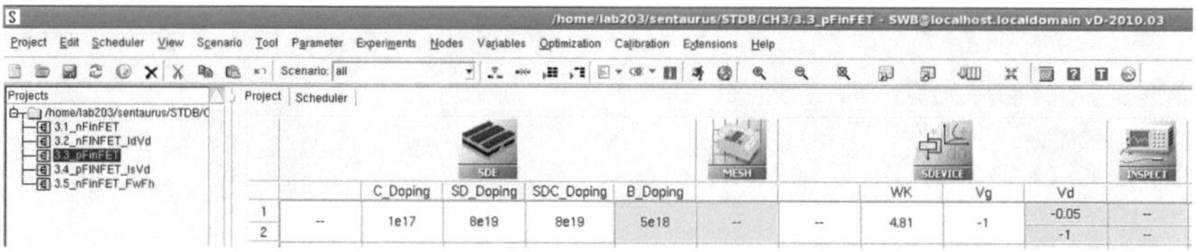

Fig. 3.35 Required simulation tools are shown in the workbench of pFinFET simulation

The SDE – devise_dvs.cmd, SDEVICE – dessis_des.cmd, and INSPECT – inspect_inc.cmd are as shown below:

1. SDE -- devise_dvs.cmd

; ----- parameter -----;

(define nm 1e-3)

(define Fw 5)

(define Fh 5)

(define Lg 15)

(define LSDC 15)

(define LSD 15)

(define Tox 0.5)

(define x1 LSDC)

(define x2 (+ x1 LSD))

(define x3 (+ x2 Lg))

(define x4 (+ x3 LSD))

(define x5 (+ x4 LSDC))

(define y1 Fw)

(define y2 (+ y1 Tox))

(define y3 (+ y2 10))

(define z1 Fh)

(define z2 (+ z1 Tox))

(define C_Doping @C_Doping@)

;(define C_Doping 1e11)

(define SD_Doping @SD_Doping@)

(define SDC_Doping @SDC_Doping@)

;(define B_Doping 1e15)

(define B_Doping @B_Doping@)

; ----- Structure -----;

"ABA"

;--- Source contact and Source ---;

(sdegeo:create-cuboid (position 0 0 0) (position x1 y1 z1) "Silicon" "SourceC")

(sdegeo:create-cuboid (position x1 0 0) (position x2 y1 z1) "Silicon" "Source")

;--- Gate oxide ---;

(sdegeo:create-cuboid (position x2 (- Tox) 0) (position x3 y2 z2) "SiO2" "Gateoxide")

;--- Channel ---;

(sdegeo:create-cuboid (position x2 0 0) (position x3 y1 z1) "Silicon" "Channel")

;--- Drain contact and Drain---;

(sdegeo:create-cuboid (position x3 0 0) (position x4 y1 z1) "Silicon" "Drain")

(sdegeo:create-cuboid (position x4 0 0) (position x5 y1 z1) "Silicon" "DrainC")

;--- Buried oxide ---;

(sdegeo:create-cuboid (position 0 (- 10) (- 20)) (position x5 y3 0) "SiO2" "Box")

"ABA"

;--- Si Body ---;

(sdegeo:create-cuboid (position 0 0 (- 20)) (position x5 y1 0) "Silicon" "Body")

; ----- Contact -----;

;----- Source -----;

(sdegeo:define-contact-set "S" 4.0 (color:rgb 1.0 0.0 0.0) "##")

(sdegeo:set-current-contact-set "S")

(sdegeo:set-contact-faces (find-face-id (position 1 1 z1)))

;----- Drain -----;

(sdegeo:define-contact-set "D" 4.0 (color:rgb 1.0 0.0 0.0) "##")

(sdegeo:set-current-contact-set "D")

(sdegeo:set-contact-faces (find-face-id (position (+ x4 1) 1 z1)))

;----- Front Gate -----;

(sdegeo:define-contact-set "G" 4.0 (color:rgb 1.0 0.0 0.0) "||")

(sdegeo:set-current-contact-set "G")

(sdegeo:set-contact-faces (find-face-id (position (+ x2 1) (- Tox) 1)))

;----- Top Gate -----;

(sdegeo:define-contact-set "G" 4.0 (color:rgb 1.0 0.0 0.0) "||")

(sdegeo:set-current-contact-set "G")

(sdegeo:set-contact-faces (find-face-id (position (+ x2 1) 1 z2)))

;----- Back Gate -----;

(sdegeo:define-contact-set "G" 4.0 (color:rgb 1.0 0.0 0.0) "||")

(sdegeo:set-current-contact-set "G")

(sdegeo:set-contact-faces (find-face-id (position (+ x2 1) y2 1)))

;----- Body -----;

(sdegeo:define-contact-set "B" 4.0 (color:rgb 1.0 0.0 0.0) "##")

(sdegeo:set-current-contact-set "B")

(sdegeo:set-contact-faces (find-face-id (position (* 0.5 x5) (* 0.5 y1) (- 20))))

; ----- Doping-----;

;----- Channel -----;

(sdedr:define-constant-profile "dopedC" "ArsenicActiveConcentration" C_Doping)

(sdedr:define-constant-profile-region "RegionC" "dopedC" "Channel")

```
;----- Source -----;

(sdedr:define-constant-profile "dopedS" "BoronActiveConcentration" SD_Doping )

(sdedr:define-constant-profile-region  "RegionS" "dopedS" "Source" )

(sdedr:define-constant-profile "dopedSC" "BoronActiveConcentration" SDC_Doping )
(sdedr:define-constant-profile-region  "RegionSC" "dopedSC" "SourceC" )

;----- Drain ------;

(sdedr:define-constant-profile "dopedD" "BoronActiveConcentration" SD_Doping )

(sdedr:define-constant-profile-region  "RegionD" "dopedD" "Drain" )

(sdedr:define-constant-profile "dopedDC" "BoronActiveConcentration" SDC_Doping )

(sdedr:define-constant-profile-region  "RegionDC" "dopedDC" "DrainC" )

;----- Si Body -----;

(sdedr:define-constant-profile "dopedB" "ArsenicActiveConcentration" B_Doping )

(sdedr:define-constant-profile-region  "RegionB" "dopedB" "Body" )

; ----- Mesh -----;

;--- AllMesh ---;

(sdedr:define-refinement-size "Cha_Mesh" 6 6 6 3 3 3)

(sdedr:define-refinement-material "channel_RF" "Cha_Mesh" "Silicon" )

;--- ChannelMesh ---;

(sdedr:define-refinement-window "multiboxChannel" "Cuboid"

(position x1 0 0)    (position x4 y1 z1))

(sdedr:define-multibox-size "multiboxSizeChannel"  2 2 2 1 1 1)
```

(sdedr:define-multibox-placement "multiboxPlacementChannel" "multiboxSizeChannel" "multiboxChannel")

(sdedr:define-refinement-function "multiboxPlacementChannel" "DopingConcentration" "MaxTransDiff" 1)

; -----Save BND and CMD and rescale to nm -----;

(sde:assign-material-and-region-names (get-body-list))

(sdeio:save-tdr-bnd (get-body-list) "n@node@_nm.tdr")

(sdedr:write-scaled-cmd-file "n@node@_msh.cmd" nm)

(define sde:scale-tdr-bnd
(lambda (tdrin sf tdrout)

  (sde:clear)

  (sdegeo:set-default-boolean "XX")

  (sdeio:read-tdr-bnd tdrin)

  (entity:scale (get-body-list) sf)

  (sdeio:save-tdr-bnd (get-body-list) tdrout)

)

)

(sde:scale-tdr-bnd "n@node@_nm.tdr" nm "n@node@_bnd.tdr")
; ----- END -----;

2. SDEVICE -- dessis_des.cmd

```
File{

    Grid="@tdr@"

    Plot="@tdrdat@"

    Current="@plot@"

    Output="@log@"

}

  Electrode {

   { name="S"    Voltage=0.0 }

   { name="D"     Voltage=0.0 }

   { name="G"       Voltage=0 WorkFunction=@WK@}

   { name="B"     Voltage=0.0 }

  }

Physics{

    Mobility( DopingDep HighFieldSaturation Enormal )

    EffectiveIntrinsicDensity( OldSlotboom )

    Recombination( SRH(DopingDep) )

    }

Math{

  Extrapolate

  Derivatives
    * Avalderivatives

    RelErrControl

    Digits=5

    ErRef(electron)=1.e10
```

```
ErRef(hole)=1.e10

Notdamped=50

Iterations=20

*Newdiscretization

Directcurrent

Method=ParDiSo

Parallel= 2
 *-VoronoiFaceBoxMethod

 NaturalBoxMethod

}
Plot{

 eDensity hDensity

 eCurrent hCurrent

 TotalCurrent/Vector eCurrent/Vector hCurrent/Vector

 eMobility hMobility

 eVelocity hVelocity

 eEnormal hEnormal
ElectricField/Vector Potential SpaceCharge

 eQuasiFermi hQuasiFermi

 Potential Doping SpaceCharge

 SRH Auger

 AvalancheGeneration

 DonorConcentration AcceptorConcentration

 Doping
 eGradQuasiFermi/Vector hGradQuasiFermi/Vector

 eEparallel hEparalllel
```

BandGap

BandGapNarrowing

Affinity

  ConductionBand ValenceBand

  }

Solve{

NewCurrentFile=""

Coupled(Iterations=100){ Poisson }

Coupled(Iterations=100){ Poisson Electron Hole }

Coupled{ Poisson Electron Hole }

Quasistationary(

InitialStep=0.01 Increment=1.35

MinStep=1e-5 MaxStep=0.2

Goal{ Name="D" Voltage= @Vd@ }

){ Coupled{ Poisson Electron Hole} }

Quasistationary(

InitialStep=1e-3 Increment=1.35

MinStep=1e-5 MaxStep=0.05

Goal{ Name="G" Voltage= @Vg@ }

){ Coupled{ Poisson Electron Hole } }

}

----- END -----

3. INSPECT -- inspect_inc.cmd

```
#-------------------------------------------------------------------------#

#        Script file designed to compute  :        #

#        * The threshold voltage        : VT   #

#        * The transconductance          : gm        #

#-------------------------------------------------------------------------#

if { ! [catch {open n@previous@_ins.log w} log_file] } {

        set fileId stdout
}

puts $log_file " "

puts $log_file "           ------------------------------------ "

puts $log_file "           Values of the extracted Parameters : "

puts $log_file "           ------------------------------------ "

puts $log_file " "

puts $log_file " "

set  DATE  [ exec  date ]

set  WORK  [ exec pwd  ]

puts $log_file " Date     : $DATE "

puts $log_file " Directory : $WORK "

puts $log_file " "

puts $log_file " "

#              idvgs=y(x) ;  vgsvgs=x(x) ;    #

set out_file n@previous@_des
proj_load "${out_file}.plt"

# ----------------------------------------------------------------- #

# I)  VT = Xintercept(maxslope(ID[VGS]))  or  VT = VGS( IDS= 100 nA/um ) #
```

```
# ---------------------------------------------------------------------- #

cv_create    idvgs   "${out_file} G OuterVoltage" "${out_file} S TotalCurrent"

cv_create    vdsvgs  "${out_file} G OuterVoltage" "${out_file} S OuterVoltage"

#................................................... #

# 1) VT extracted as the intersection point with the X axis at the point #

#    where the id(vgs) slope reaches its maxmimum :    #

#................................................... #

set VT1   [ f_VT1 idvgs ]

#..........................................         #

# 2) Printing of the whole set of extracted values (std output) :    #

#..........................................         #

puts $log_file "Threshold   voltage VT1 = $VT1 Volts"

puts $log_file " "

#................................................... #

# 3) Initialization and display of curves on the main Inspect screen  :  #

# ................................................... #

cv_display    idvgs

cv_lineStyle idvgs  solid

cv_lineColor idvgs  red

# ---------------------------------------------------------------------- #

# II)                    gm =  maxslope((ID[VGS])      #

# ---------------------------------------------------------------------- #

set gm    [ f_gm idvgs ]

puts $log_file " "

puts $log_file "Transconductance gm                    = $gm  A/V"
```

```
puts $log_file " "

set ioff [ cv_compute "vecmin(<idvgs>)" A A A A ]

puts $log_file " "

puts $log_file "Current ioff      = $ioff  A"

puts $log_file " "

set isat [ cv_compute "vecmax(<idvgs>)" A A A A ]

puts $log_file " "

puts $log_file "Current isat      = $isat  A"

puts $log_file " "

set rout [ cv_compute "Rout(<idvgs>)" A A A A ]

puts $log_file " "

puts $log_file "Resistant rout      = $rout  A"

puts $log_file " "

cv_createWithFormula logcurve "log10(<idvgs>)" A A A A

cv_createWithFormula difflog "(-1)*diff(<logcurve>)" A A A A

set sslop [ cv_compute "1/vecmax(<difflog>)" A A A A ]

puts $log_file " "

puts $log_file "sub solp      = $sslop  A/V"

puts $log_file " "

### Puting into Family Table #####

ft_scalar VT $VT1

ft_scalar gmax $gm

ft_scalar ioff $ioff

ft_scalar isat $isat

ft_scalar sslop $sslop
```

ft_scalar rout $rout

close $log_file

----- END -----#

The I_d–V_g curve of the simulation result is as shown in Fig. 3.36, in which the important parameters are SS at around 64 mV/dec., V_{th} at around −0.3 V, I_{sat} at 1.29×10^{-5} A, and I_{off} at 2.93×10^{-12} A as shown in Fig. 3.37. The structural channel mesh, the electron concentration distributions of 3D and 2D structures, electric field distributions, electric potential distributions, and the energy band diagrams along the channel direction are as shown in Fig. 3.38, 3.39, 3.40, 3.41, 3.42, 3.43, 3.44 and 3.45 with the conditions of L_g = 15 nm, V_d = −1 V, and V_g = −1 V, respectively.

Example 3.4 Comparison of different F_h (Fin height) with L_g = 10 nm nFinFET

The following three main program code files are based on **Synopsys Sentaurus TCAD 2014 version.**

We use gate length L_g = 10 nm nFinFET with F_w = 5 nm and F_h = 5, 10, 15, 20, 25, 30 and 35 nm at V_{dd} = 0.7 V are fir simulation (Figs. 3.46, 3.47 and 3.48).

3D - pFinFET

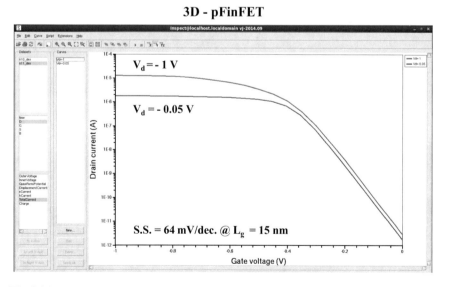

Fig. 3.36 I_d–V_g curve of 3D pFinFET simulation

		VT	gmax	ioff	isat	sslop	rout
1	--	-0.3110238343247096	384579140556	1.7512775867682e-12	1.81593870786853e-06	0.0644208330893465	9.3745993886
2	--	-0.2979128544605598	55163603945	2.9284554147177e-12	1.28610734168877e-05	0.06405853158951812	82339014707

Fig. 3.37 Electrical property parameters of 3D pFinFET simulation

3D – pFinFET Mesh

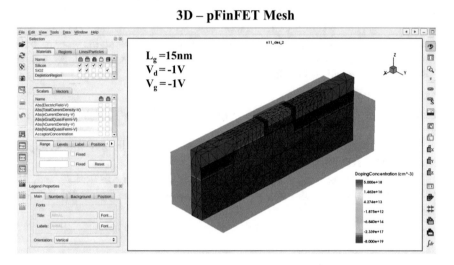

Fig. 3.38 Mesh diagram of 3D pFinFET simulation

3D – Hole concentration

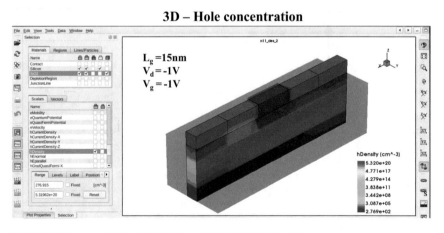

Fig. 3.39 Hole concentration distribution of 3D pFinFET simulation

2D – Hole concentration

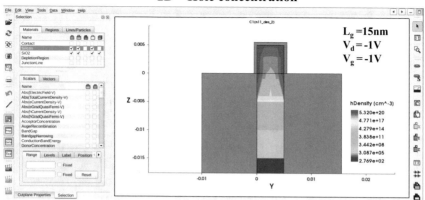

Fig. 3.40 Hole concentration distribution of 2D pFinFET simulation

3D – Electric Field

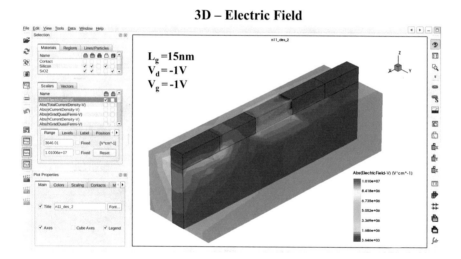

Fig. 3.41 Electric field distribution of 3D pFinFET simulation

2D – Electric Field

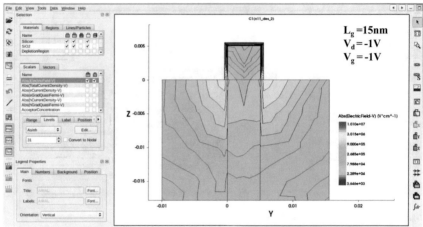

Fig. 3.42 Electric field distribution of 2D cross-sectional plot of pFinFET simulation

3D – Electrostatic Potential

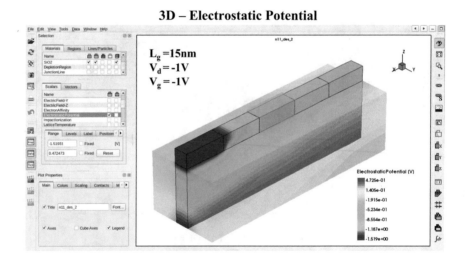

Fig. 3.43 Electric potential distribution of 3D pFinFET simulation

2D – Electrostatic Potential

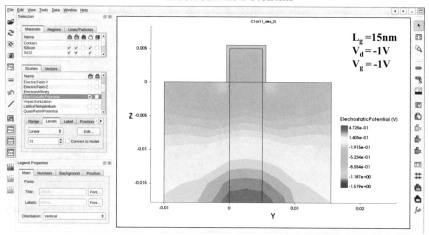

Fig. 3.44 Electric potential distribution of 2D cross-sectional plot of pFinFET simulation

Band Diagram

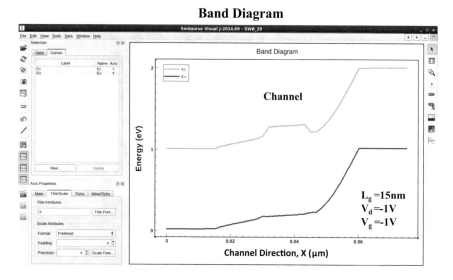

Fig. 3.45 Energy band diagram of 3D pFinFET simulation

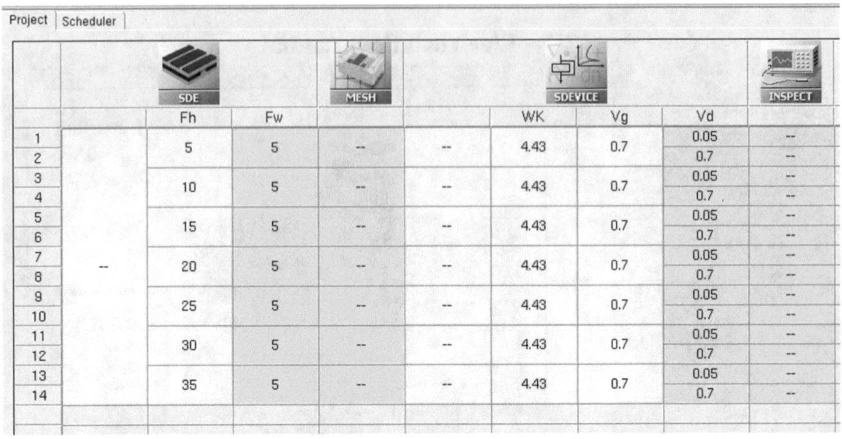

Project	Scheduler							
		SDE		MESH		SDEVICE		INSPECT
		Fh	Fw			WK	Vg	Vd
1		5	5	--	--	4.43	0.7	0.05
2								0.7
3		10	5	--	--	4.43	0.7	0.05
4								0.7
5		15	5	--	--	4.43	0.7	0.05
6								0.7
7	--	20	5	--	--	4.43	0.7	0.05
8								0.7
9		25	5	--	--	4.43	0.7	0.05
10								0.7
11		30	5	--	--	4.43	0.7	0.05
12								0.7
13		35	5	--	--	4.43	0.7	0.05
14								0.7

Fig. 3.46 Required simulation tools are shown in the workbench of L_g = 10 nm nFinFET with different F_h simulations and V_{dd} = 0.7 V

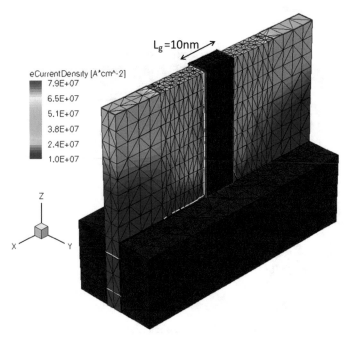

Fig. 3.47 Electron current density and mesh plots of L_g = 10 nm nFinFET with F_w = 5 nm and F_h = 35 m at V_{dd} = 0.7 V

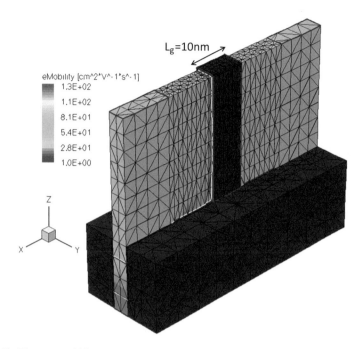

Fig. 3.48 Electron mobility and mesh plots of $L_g = 10$ nm nFinFET with $F_w = 5$ nm and $F_h = 35$ nm at $V_{dd} = 0.7$ V. The unstrained Si channel maximum mobility is around 130 cm^2/Vs

The SDE – devise_dvs.cmd, SDEVICE – dessis_des.cmd, and INSPECT – inspect_inc.cmd are as shown below:

1. SDE -- devise_dvs.cmd

; --------- parameter ------------;

(define nm 1e-3)

(define Fw @Fw@)

(define Fh @Fh@)

(define Lg 10)

(define LSDC 15)

(define LSD 15)

(define Tox 0.5)

(define x1 LSDC)

(define x2 (+ x1 LSD))

(define x3 (+ x2 Lg))

(define x4 (+ x3 LSD))

(define x5 (+ x4 LSDC))

(define y1 Fw)

(define y2 (+ y1 Tox))

(define y3 (+ y2 10))

(define z1 Fh)

(define z2 (+ z1 Tox))

(define C_Doping 1e17)

(define SD_Doping 8e19)

(define SDC_Doping 8e19)

(define B_Doping 5e18)

; --------- Structure ---------;

"ABA"

```
;--- Source contact and Source ---;

(sdegeo:create-cuboid (position 0 0 0 ) (position x1 y1 z1 ) "Silicon" "SourceC")

(sdegeo:create-cuboid (position x1 0 0 ) (position x2 y1 z1) "Silicon" "Source")

;--- Gate oxide ---;

(sdegeo:create-cuboid (position x2 (- Tox) 0 ) (position  x3 y2 z2 ) "SiO2" "Gateoxide")

;--- Channel ---;

(sdegeo:create-cuboid (position x2 0 0 ) (position x3 y1 z1 ) "Silicon" "Channel")

;--- Drain contact and Drain---;

(sdegeo:create-cuboid (position x3 0 0 ) (position x4 y1 z1 ) "Silicon" "Drain")

(sdegeo:create-cuboid (position x4 0 0 ) (position x5 y1 z1 ) "Silicon" "DrainC")

;--- Buried oxide ---;

(sdegeo:create-cuboid (position 0 (- 10) (- 20) ) (position x5 y3 0 ) "SiO2" "Box")

;--- Si Body ---;

(sdegeo:create-cuboid (position 0 0 (- 20) ) (position x5 y1 0 ) "Silicon" "Body")

; ------------- Contact --------------;

;----- Source -----;

(sdegeo:define-contact-set "S" 4.0  (color:rgb 1.0 0.0 0.0 ) "##" )

(sdegeo:set-current-contact-set "S")

(sdegeo:set-contact-faces (find-face-id (position 1 1 z1)))

;----- Drain -----;

(sdegeo:define-contact-set "D" 4.0  (color:rgb 1.0 0.0 0.0 ) "##" )

(sdegeo:set-current-contact-set "D")

(sdegeo:set-contact-faces (find-face-id (position (+ x4 1) 1 z1 )))

;----- Front Gate -----;

(sdegeo:define-contact-set "G" 4.0  (color:rgb 1.0 0.0 0.0 ) "||" )
```

```
(sdegeo:set-current-contact-set "G")

(sdegeo:set-contact-faces (find-face-id (position (+ x2 1) (- Tox) 1 )))

;----- Top Gate -----;

(sdegeo:define-contact-set "G" 4.0  (color:rgb 1.0 0.0 0.0 ) "||" )

(sdegeo:set-current-contact-set "G")

(sdegeo:set-contact-faces (find-face-id (position (+ x2 1) 1  z2 )))

;----- Back Gate -----;

(sdegeo:define-contact-set "G" 4.0  (color:rgb 1.0 0.0 0.0 ) "||" )

(sdegeo:set-current-contact-set "G")

(sdegeo:set-contact-faces (find-face-id (position (+ x2 1) y2 1 )))

;----- Body -----;

(sdegeo:define-contact-set "B" 4.0  (color:rgb 1.0 0.0 0.0 ) "##" )

(sdegeo:set-current-contact-set "B")

(sdegeo:set-contact-faces (find-face-id (position (* 0.5 x5) (* 0.5 y1) (- 20) )))

; -----Doping -----;

;----- Channel -----;

(sdedr:define-constant-profile "dopedC" "BoronActiveConcentration" C_Doping )

(sdedr:define-constant-profile-region  "RegionC" "dopedC" "Channel" )

;----- Source -----;

(sdedr:define-constant-profile "dopedS" "ArsenicActiveConcentration" SD_Doping )

(sdedr:define-constant-profile-region  "RegionS" "dopedS" "Source" )

(sdedr:define-constant-profile "dopedSC" "ArsenicActiveConcentration" SDC_Doping )

(sdedr:define-constant-profile-region  "RegionSC" "dopedSC" "SourceC" )

;----- Drain ------;

(sdedr:define-constant-profile "dopedD" "ArsenicActiveConcentration" SD_Doping )
```

(sdedr:define-constant-profile-region "RegionD" "dopedD" "Drain")

(sdedr:define-constant-profile "dopedDC" "ArsenicActiveConcentration" SDC_Doping)

(sdedr:define-constant-profile-region "RegionDC" "dopedDC" "DrainC")

;----- Si Body -----;

(sdedr:define-constant-profile "dopedB" "BoronActiveConcentration" B_Doping)

(sdedr:define-constant-profile-region "RegionB" "dopedB" "Body")

; ------------- Mesh -------------;

;--- AllMesh ---;

(sdedr:define-refinement-size "Cha_Mesh" 5 5 5 1 1 1)

(sdedr:define-refinement-material "channel_RF" "Cha_Mesh" "Silicon")

;--- ChannelMesh ---;

(sdedr:define-refinement-window "multiboxChannel" "Cuboid"

(position x1 0 0)   (position x4 y1 z1))

(sdedr:define-multibox-size "multiboxSizeChannel" 2 2 2 2 2 2)

(sdedr:define-multibox-placement "multiboxPlacementChannel" "multiboxSizeChannel"
"multiboxChannel")

(sdedr:define-refinement-function "multiboxPlacementChannel" "DopingConcentration"
"MaxTransDiff" 1)

; ----- Save BND and CMD and rescale to nm-----;

(sde:assign-material-and-region-names (get-body-list))

(sdeio:save-tdr-bnd (get-body-list) "n@node@_nm.tdr")

(sdedr:write-scaled-cmd-file "n@node@_msh.cmd" nm)

(define sde:scale-tdr-bnd

 (lambda (tdrin sf tdrout)

  (sde:clear)

  (sdegeo:set-default-boolean "XX")

(sdeio:read-tdr-bnd tdrin)

(entity:scale (get-body-list) sf)

(sdeio:save-tdr-bnd (get-body-list) tdrout)

))

(sde:scale-tdr-bnd "n@node@_nm.tdr" nm "n@node@_bnd.tdr")

; ------------------ END ------------------------;

2. SDEVICE -- dessis_des.cmd

File{

    Grid="@tdr@"

    Plot="@tdrdat@"

    Current="@plot@"

    Output="@log@"

}

  Electrode {

   { name="S"   Voltage=0.0 }

   { name="D"   Voltage=0.0 }

   { name="G"   Voltage=0 WorkFunction=@WK@}

   { name="B"   Voltage=0.0 }

  }

```
Physics{

   Mobility( DopingDep HighFieldSaturation Enormal )

   EffectiveIntrinsicDensity( OldSlotboom )

   Recombination( SRH(DopingDep) )

   }

Math{

  Extrapolate

  Derivatives

  * Avalderivatives

  RelErrControl

  Digits=5

  ErRef(electron)=1.e10
  ErRef(hole)=1.e10

  Notdamped=50

  Iterations=20

  *Newdiscretization

  Directcurrent

  Method=ParDiSo

  Parallel= 2

  *-VoronoiFaceBoxMethod

  NaturalBoxMethod

}
```

Plot{

  eDensity hDensity

  eCurrent hCurrent

  TotalCurrent/Vector eCurrent/Vector hCurrent/Vector

  eMobility hMobility

  eVelocity hVelocity

  eEnormal hEnormal

  ElectricField/Vector Potential SpaceCharge

  eQuasiFermi hQuasiFermi

    Potential Doping SpaceCharge

    SRH Auger

    AvalancheGeneration

    DonorConcentration AcceptorConcentration

    Doping
    eGradQuasiFermi/Vector hGradQuasiFermi/Vector
    eEparallel hEparalllel

    BandGap

    BandGapNarrowing

    Affinity

    ConductionBand ValenceBand

  }

```
Solve{

NewCurrentFile=""

Coupled(Iterations=100){ Poisson }

Coupled(Iterations=100){ Poisson Electron Hole }

Coupled{ Poisson Electron Hole }

Quasistationary(
InitialStep=0.01 Increment=1.35

MinStep=1e-5 MaxStep=0.2

Goal{ Name="D" Voltage= @Vd@ }

){ Coupled{ Poisson Electron Hole } }

Quasistationary(

InitialStep=1e-3 Increment=1.35

MinStep=1e-5 MaxStep=0.05

Goal{ Name="G" Voltage= @Vg@ }

){ Coupled{ Poisson Electron Hole } }

}
```

* ---------------------- END ----------------------*

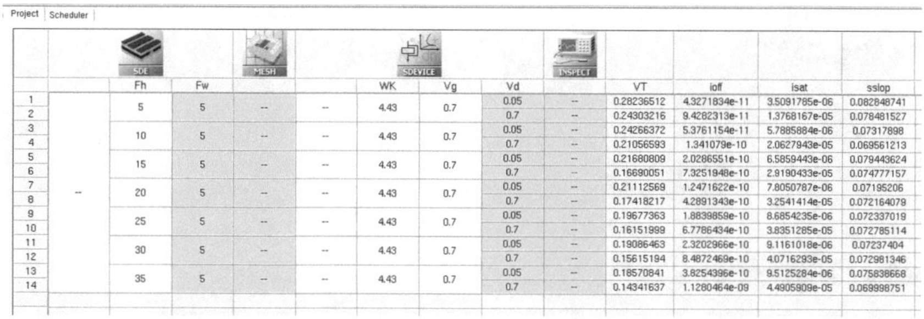

	Fh	Fw		WK	Vg	Vd		VT	ioff	isat	sslop
1	5	5	--	4.43	0.7	0.05	--	0.28236512	4.3271834e-11	3.5091765e-06	0.082648741
2						0.7	--	0.24303216	9.4282313e-11	1.3768167e-05	0.078481527
3	10	5	--	4.43	0.7	0.05	--	0.24266372	5.3761154e-11	5.7885884e-06	0.07317898
4						0.7	--	0.21056593	1.341079e-10	2.0627943e-05	0.069561213
5	15	5	--	4.43	0.7	0.05	--	0.21680809	2.0286551e-10	6.5859443e-06	0.079443624
6						0.7	--	0.16690051	7.3251948e-10	2.9190433e-05	0.074777157
7	20	5	--	4.43	0.7	0.05	--	0.21112569	1.2471622e-10	7.8050787e-06	0.07195206
8						0.7	--	0.17418217	4.2891343e-10	3.2541414e-05	0.072164079
9	25	5	--	4.43	0.7	0.05	--	0.19677363	1.8839858e-10	8.6854235e-06	0.072337019
10						0.7	--	0.16151999	6.7786434e-10	3.8351285e-05	0.072785114
11	30	5	--	4.43	0.7	0.05	--	0.19086463	2.3202966e-10	9.1161018e-06	0.07237404
12						0.7	--	0.15615194	8.4872469e-10	4.0716293e-05	0.072981346
13	35	5	--	4.43	0.7	0.05	--	0.18570841	3.8254396e-10	9.5125284e-06	0.075838668
14						0.7	--	0.14341637	1.1280464e-09	4.4905909e-05	0.069998751

Fig. 3.49 Results electric properties of $L_g = 10$ nm nFinFET of $F_w = 5$ nm and different F_h at $V_{dd} = 0.7$ V. The important parameters V_{th}, I_{off}, I_{sat}, and SS are shown

3. INSPECT -- inspect_inc.cmd

```
#-----------------------------------------------------------------------#
#       Script file designed to compute  :          #
#        * The threshold voltage       : VT   #                #              #
#        * The transconductance        : gm          #
#-----------------------------------------------------------------------#
if { ! [catch {open n@previous@_ins.log w} log_file] } {
        set fileId stdout
}
puts $log_file " "
puts $log_file "            ------------------------------------ "
puts $log_file "            Values of the extracted Parameters : "
puts $log_file "            ------------------------------------ "
puts $log_file " "
puts $log_file " "
set  DATE  [ exec  date ]
set  WORK  [ exec pwd  ]
puts $log_file " Date     : $DATE "
puts $log_file " Directory : $WORK "
puts $log_file " "
puts $log_file " "
#                                                       #
#            idvgs=y(x) ;  vgsvgs=x(x) ; #
#                                                       #
set out_file n@previous@_des
proj_load "${out_file}.plt"
```

```
# ------------------------------------------------------------------- #
# I)  VT = Xintercept(maxslope(ID[VGS]))  or  VT = VGS( IDS= 0.1 ua/um ) #
# ------------------------------------------------------------------- #
cv_create   idvgs  "${out_file} G OuterVoltage" "${out_file} D TotalCurrent"
cv_create   vdsvgs "${out_file} G OuterVoltage" "${out_file} D OuterVoltage"
#..................................................... #
# 1) VT extracted as the intersection point with the X axis at the point #
#   where the id(vgs) slope reaches its maxmimum : #
#..................................................... #
set VT1   [ f_VT1 idvgs ]
#...........................................          #
# 2) Printing of the whole set of extracted values (std output) :   #
#...........................................          #
puts $log_file "Threshold   voltage VT1  = $VT1 Volts"
puts $log_file " "
#.................................................. #
# 3) Initialization and display of curves on the main Inspect screen  : #
# .................................................. #
cv_display   idvgs
cv_lineStyle idvgs  solid
cv_lineColor idvgs  red
# ------------------------------------------------------------------- #
# II)                    gm =  maxslope((ID[VGS])         #
# ------------------------------------------------------------------- #
set gm    [ f_gm idvgs ]
puts $log_file " "
```

puts $log_file "Transconductance gm   = $gm A/V"

puts $log_file " "

set ioff [cv_compute "vecmin(<idvgs>)" A A A A]

puts $log_file " "

puts $log_file "Current ioff   = $ioff A"

puts $log_file " "

set isat [cv_compute "vecmax(<idvgs>)" A A A A]

puts $log_file " "

puts $log_file "Current isat   = $isat A"

puts $log_file " "

set rout [cv_compute "Rout(<idvgs>)" A A A A]

puts $log_file " "

puts $log_file "Resistant rout   = $rout A"

puts $log_file " "

cv_createWithFormula logcurve "log10(<idvgs>)" A A A A

cv_createWithFormula difflog "diff(<logcurve>)" A A A A

set sslop [cv_compute "1/vecmax(<difflog>)" A A A A]

puts $log_file " "

puts $log_file "sub solp   = $sslop A/V"

puts $log_file " "

Puting into Family Table

ft_scalar VT $VT1

ft_scalar gmax $gm

ft_scalar ioff $ioff

ft_scalar isat $isat

ft_scalar sslop $sslop

ft_scalar rout $rout

close $log_file

------------------ END -------------------#

The important key performance index (KPI), V_{th}, I_{off}, I_{on} and SS, I_{sat}, with respect to different F_h is as shown in Fig. 3.49 . The I_d–V_g of simulation result is as shown in Fig. 3.50. It is shown in the figure that smaller F_h will lead to smaller I_{off} and SS, indicating better gate control capability. Again, it has been mentioned previously that the smaller Natural length will be preferred. In general, $L_g > 5$–10 times of λ. Thus, smaller F_h will lead to smaller value of natural length.

Example 3.5 $Si_{1-x}Ge_x$ pFinFET with $L_g = 15$ nm and HfO_2 as gate insulator Material selection for sub-7-nm technology node:

(a) **Silicon**

Silicon is the most important semiconductor material of the world. It is fairly abundant in nature, easy to be process, and equipped with characteristics of nFET and pFET with excellent stability and very low cost. These characteristics have made silicon the favorite material in semiconductor industry until 10-nm nodes or beyond.

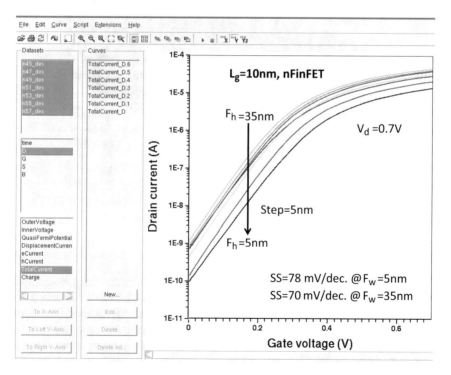

Fig. 3.50 Comparison among I_d–V_g curves of $L_g = 10$ nm nFinFET with $F_w = 5$ nm and different F_h at $V_{dd} = 0.7$ V. The higher F_h will lead to higher I_{sat} and also higher I_{off}

	THfO2	C_Doping	SD_Doping	SDC_Doping	B_Doping			WK	xF	Vg	Vd	
1									0	-1	-1.0	--
2									0.1	-1	-1.0	--
3									0.2	-1	-1.0	--
4									0.3	-1	-1.0	--
5									0.4	-1	-1.0	--
6	--	3	1e17	8e19	8e19	5e18		--	4.6	0.5	-1	-1.0	--
7									0.6	-1	-1.0	--
8									0.7	-1	-1.0	--
9									0.8	-1	-1.0	--
10									0.9	-1	-1.0	--
11									1.0	-1	-1.0	--

Fig. 3.51 SWB shows **Si$_{1-x}$Ge$_x$ pFinFET** with L_g = 15 nm, and set **Mole-Fraction** x as xF from 0 to 1.0 as step 0.1, where silicongermanium $(x = 0)$ = silicon and silicongermanium $(x = 1)$ = germanium

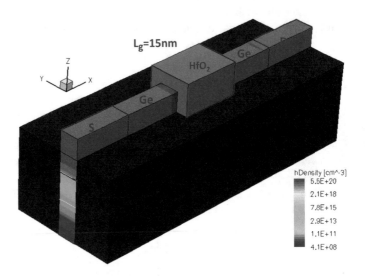

Fig. 3.52 SWB shows **Si$_{1-x}$Ge$_x$ pFinFET** with HfO$_2$ of L_g = 15 nm

(b) **Germanium**

Germanium is also equipped with many advantages. For example, it is equipped with higher electron mobility and hole mobility than silicon of Table 3.4, and it can be used for high-frequency field. These advantages plus the similar fabrication process to silicon have made Germanium a possible candidate for applications at sub-10-nm node. The Si$_{1-x}$Ge$_x$ material has higher mobility and is compatible to current Si-based FinFET process. Therefore, for Example 3.5, we study the Si$_{1-x}$Ge$_x$ FinFET by using high-k material HfO$_2$ of 3 nm. The Si$_{1-x}$Ge$_x$ Ge molecular fraction X changes from 0 to 1, step 0.1, which is shown in Fig. 3.51.

In Sentaurus Device Mole-Fraction Materials section (Sentaurus™ Device User Guide J-2014.09):

xF	Vg	Vd		VT	gmax	ioff	isat	sslop	rout
0	-1	-1.0	--	-0.4938222	-3.6937223e-13	1.1160346e-14	1.2243531e-05	0.069544491	-31.671598
0.1	-1	-1.0	--	-0.41493771	-4.8228433e-12	1.4972663e-13	1.4675701e-05	0.071471351	-30.827772
0.2	-1	-1.0	--	-0.34249562	-6.006689e-11	1.8640041e-12	1.6914775e-05	0.072093258	-30.062028
0.3	-1	-1.0	--	-0.26385503	-7.7025792e-10	2.448522e-11	1.8988526e-05	0.073056376	-29.410949
0.4	-1	-1.0	--	-0.19118792	-8.2828247e-09	2.6704458e-10	2.0724202e-05	0.074087028	-28.851083
0.5	-1	-1.0	--	-0.12445039	-6.8754146e-08	2.259656e-09	2.2103033e-05	0.075677317	-28.377826
0.6	-1	-1.0	--	-0.05990411	-5.1745704e-07	1.765033e-08	2.3222427e-05	0.078541751	-27.642269
0.7	-1	-1.0	--	-0.033240691	-1.1351275e-06	3.92624e-08	2.3688583e-05	0.080788878	-27.898177
0.8	-1	-1.0	--	-0.016813742	-1.7723963e-06	6.2823925e-08	2.3997885e-05	0.082769638	-27.639145
0.9	-1	-1.0	--	-0.0066891199	-2.306194e-06	8.336333e-08	2.4223046e-05	0.084378934	-27.286294
1.0	-1	-1.0	--		-2.8815586e-06	1.0618165e-07	2.441992e-05	0.085993387	-26.97725

Fig. 3.53 Results electric properties of $Si_{1-x}Ge_x$ pFinFET with $L_g = 15$ nm, and set **Mole-Fraction** x as xF from 0 to 1.0 as step 0.1, where silicongermanium ($x = 0$) = silicon and silicongermanium ($x = 1$) = germanium. The important parameters V_{th}, I_{off}, I_{sat}, and SS are shown

Table 3.4 Electron and hole mobility of important semiconductor materials

Material $(cm^2 \ V^{-1} \ s^{-1})$	Si	Ge	GaAs	InGaAs	InAs
Mobility (electrons)	1350	3600	8000	11,200	30,000
Mobility (holes)	480	1800	300	300	450

Fig. 3.54 Energy band diagram of $L_g = 15$ nm Ge pFinFET using $x = 1$ molecular fraction value of $Si_{1-x}Ge_x$ from Fig. 5.53

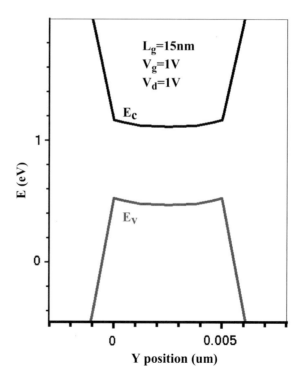

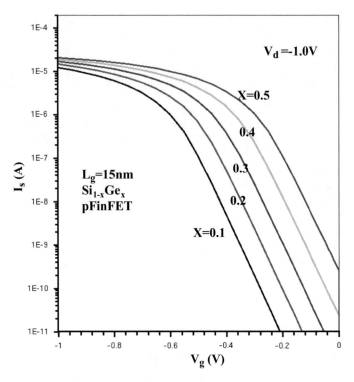

Fig. 3.55 Extraction I_s–V_g curves from Fig. 3.53 $Si_{1-x}Ge_x$ pFinFET with L_g = 15 nm and different x fractions from 0.1 to 0.5. The higher x value has higher I_{on} and I_{off}

Fig. 3.56 Important semiconductor materials band gap and lattice constant for FinFET active channel

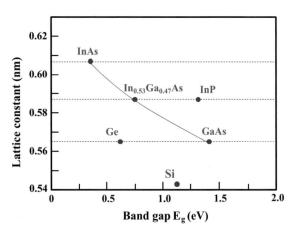

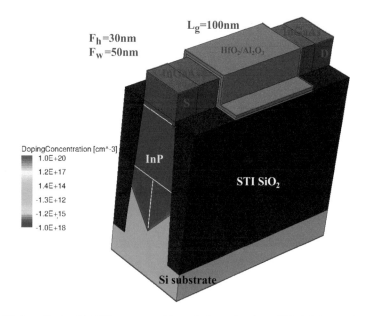

Fig. 3.57 In$_{1-x}$Ga$_x$As nFinFET structure with stacked gate insulator HfO$_2$/Al$_2$O$_3$ (*top/bottom*) of 3/3 nm

Sentaurus Device reads the file Molefraction.txt to determine mole fraction-dependent materials. The following search strategy is used to locate this file:

1. Sentaurus Device looks for Molefraction.txt in the current working directory.
2. If the environment variables STROOT and STRELEASE are defined, Sentaurus Device tries to read the file (Fig. 3.52):
   $STROOT/tcad/$STRELEASE/lib/Molefraction.txt

	Lg	Fh		xF	WK	Vg	Vd		gmax	ioff	isat	sslop	VT
1				0	4.85	1	0.05	--	0.0015283757	1.0856228e-05	0.0013688196	0.083644957	
2							1	--	0.0010794081	9.0386544e-06	0.00098613506	0.085695116	
3				0.2	4.85	1	0.05	--	0.0013640569	5.9880351e-08	0.001039399	0.066483464	0.01473008
4							1	--	0.00097625055	5.6745374e-08	0.00075248347	0.068684986	0.017677056
5				0.4	4.85	1	0.05	--	0.001216248	1.7059493e-10	0.00072674754	0.064292239	0.16914877
6							1	--	0.00086053857	1.9035406e-10	0.00053739251	0.06623239	0.18198934
7	--	100	30	0.6	4.85	1	0.05	--	0.0010704014	3.8668805e-13	0.00046010971	0.062689148	0.33482579
8							1	--	0.00074932655	4.8779555e-13	0.00034725988	0.065135476	0.341187
9				0.8	4.85	1	0.05	--	0.00085206623	7.6825921e-16	0.00024617832	0.036797612	0.4987094
10							1	--	0.00060786317	1.0851667e-15	0.00019894474	0.019141961	0.51278309
11				1.0	4.85	1	0.05	--	0.00056049591	-1.1332735e-17	9.306948e-05	1.1275788e-05	0.6654038
12							1	--	0.0005208225	-0.0661698e-17	8.4270203e-05	3.6757981e-06	0.67828826
13				0.47	4.85	1	0.05	--	0.0011704525	2.0632339e-11	0.00082780946	0.063552414	0.22969299
14							1	--	0.00081846671	2.4076427e-11	0.00046756943	0.065727354	0.2391051

Fig. 3.58 SWB shows In$_{1-x}$Ga$_x$As nFinFET with L_g = 100 nm, and set **Mole-Fraction** *x* as *x*F from 0 to 1.0 as step 0.2, and 0.47. For *x* = 0.47, In$_{1-x}$Ga$_x$As is In$_{0.53}$Ga$_{0.47}$As

3. If these previous strategies are unsuccessful, Sentaurus Device uses the built-in defaults that follow.

The default Molefraction.txt file has the following content:
 # Ge(x)Si(1-x)

SiliconGermanium (x=0) = Silicon

SiliconGermanium (x=1) = Germanium

 # Note: x can be set as a variable between 0 to 1.

Other compound semiconductors are defined as follows:
Al(x)Ga(1-x)As

AlGaAs (x=0) = GaAs

AlGaAs (x=1) = AlAs

In(1-x)Al(x)As

InAlAs (x=0) = InAs

InAlAs (x=1) = AlAs

 # In(1-x)Ga(x)As

InGaAs (x=0) = InAs

InGaAs (x=1) = GaAs

Ga(x)In(1-x)P

GaInP (x=0) = InP

GaInP (x=1) = GaP

InAs(x)P(1-x)

InAsP (x=0) = InP

InAsP (x=1) = InAs

GaAs(x)P(1-x)

GaAsP (x=0) = GaP

GaAsP (x=1) = GaAs

Hg(1-x)Cd(x)Te

HgCdTe (x=0) = HgTe

HgCdTe (x=1) = CdTe

In(1-x)Ga(x)As(y)P(1-y)

InGaAsP (x=0, y=0) = InP

InGaAsP (x=1, y=0) = GaP

InGaAsP (x=1, y=1) = GaAs

InGaAsP (x=0, y=1) = InAs

1. SDE tool

;---------- example 3.5 SiGex pFinFET with Lg=15nm and HFO2 ---------;

example 3.1 nFinFET Lg = 15 nm

;----------------------------------- parameter ---;

(define nm 1e-3)

(define Fw 5)

(define Fh 5)

(define Lg 15)

(define LSDC 15)

(define LSD 15)

(define THfO2 @THfO2@)

(define x1 LSDC)

(define x2 (+ x1 LSD))

(define x3 (+ x2 Lg))

(define x4 (+ x3 LSD))

(define x5 (+ x4 LSDC))

(define y1 Fw)

(define y2 (+ y1 THfO2))

(define y3 (+ y2 10))

(define z1 Fh)

(define z2 (+ z1 THfO2))

(define C_Doping @C_Doping@)

;(define C_Doping 1e11)

(define SD_Doping @SD_Doping@)

(define SDC_Doping @SDC_Doping@)

;(define B_Doping 1e15)

(define B_Doping @B_Doping@)

;---------------------------------- Structure --;

"ABA"

;--- Source contact and Source ---;

(sdegeo:create-cuboid (position 0 0 0) (position x1 y1 z1) "SiliconGermanium" "SourceC")

(sdegeo:create-cuboid (position x1 0 0) (position x2 y1 z1) "SiliconGermanium" "Source")

;--- Gate oxide ---;

(sdegeo:create-cuboid (position x2 (- THfO2) 0) (position x3 y2 z2) "HfO2" "Gateoxide")

;--- Channel ---;

(sdegeo:create-cuboid (position x2 0 0) (position x3 y1 z1) "SiliconGermanium" "Channel")

;--- Drain contact and Drain---;

(sdegeo:create-cuboid (position x3 0 0) (position x4 y1 z1) "SiliconGermanium" "Drain")

(sdegeo:create-cuboid (position x4 0 0) (position x5 y1 z1) "SiliconGermanium" "DrainC")

;--- Buried oxide ---;

(sdegeo:create-cuboid (position 0 (- 10) (- 20)) (position x5 y3 0) "SiO2" "Box")

"ABA"

;--- Si Body ---;

(sdegeo:create-cuboid (position 0 0 (- 20)) (position x5 y1 0) "SiliconGermanium" "Body")

;-------------------------------------- Contact --------------------------------------;

;----- Source -----;

(sdegeo:define-contact-set "S" 4.0 (color:rgb 1.0 0.0 0.0) "##")

(sdegeo:set-current-contact-set "S")

(sdegeo:set-contact-faces (find-face-id (position 1 1 z1)))

;----- Drain -----;

(sdegeo:define-contact-set "D" 4.0 (color:rgb 1.0 0.0 0.0) "##")

(sdegeo:set-current-contact-set "D")

(sdegeo:set-contact-faces (find-face-id (position (+ x4 1) 1 z1)))

;----- Front Gate -----;

(sdegeo:define-contact-set "G" 4.0 (color:rgb 1.0 0.0 0.0) "||")

(sdegeo:set-current-contact-set "G")

(sdegeo:set-contact-faces (find-face-id (position (+ x2 1) (- THfO2) 1)))

;----- Top Gate -----;

(sdegeo:define-contact-set "G" 4.0 (color:rgb 1.0 0.0 0.0) "||")

(sdegeo:set-current-contact-set "G")

(sdegeo:set-contact-faces (find-face-id (position (+ x2 1) 1 z2)))

;----- Back Gate -----;

(sdegeo:define-contact-set "G" 4.0 (color:rgb 1.0 0.0 0.0) "||")

(sdegeo:set-current-contact-set "G")

(sdegeo:set-contact-faces (find-face-id (position (+ x2 1) y2 1)))

;----- Body -----;

(sdegeo:define-contact-set "B" 4.0 (color:rgb 1.0 0.0 0.0) "##")

(sdegeo:set-current-contact-set "B")

(sdegeo:set-contact-faces (find-face-id (position (* 0.5 x5) (* 0.5 y1) (- 20))))

;-------------------------------------- Doping --------------------------------------;

;----- Channel -----;

(sdedr:define-constant-profile "dopedC" "ArsenicActiveConcentration" C_Doping)

(sdedr:define-constant-profile-region "RegionC" "dopedC" "Channel")

;----- Source -----;

(sdedr:define-constant-profile "dopedS" "BoronActiveConcentration" SD_Doping)

(sdedr:define-constant-profile-region "RegionS" "dopedS" "Source")

(sdedr:define-constant-profile "dopedSC" "BoronActiveConcentration" SDC_Doping)

(sdedr:define-constant-profile-region "RegionSC" "dopedSC" "SourceC")

;----- Drain ------;

(sdedr:define-constant-profile "dopedD" "BoronActiveConcentration" SD_Doping)

(sdedr:define-constant-profile-region "RegionD" "dopedD" "Drain")

(sdedr:define-constant-profile "dopedDC" "BoronActiveConcentration" SDC_Doping)

(sdedr:define-constant-profile-region "RegionDC" "dopedDC" "DrainC")

;----- Si Body -----;

(sdedr:define-constant-profile "dopedB" "ArsenicActiveConcentration" B_Doping)

(sdedr:define-constant-profile-region "RegionB" "dopedB" "Body")

;-------------------------------------- Mesh --------------------------------------;

```
;--- AllMesh ---;

(sdedr:define-refinement-size "Cha_Mesh" 5 5 5 1 1 1)

(sdedr:define-refinement-material "channel_RF" "Cha_Mesh" "SiliconGermanium" )

;--- ChannelMesh ---;

(sdedr:define-refinement-window "multiboxChannel" "Cuboid"

(position x1 0 0)   (position x4 y1 z1))

(sdedr:define-multibox-size "multiboxSizeChannel" 2 2 2 2 2 2)

(sdedr:define-multibox-placement "multiboxPlacementChannel" "multiboxSizeChannel"
"multiboxChannel")

(sdedr:define-refinement-function "multiboxPlacementChannel" "DopingConcentration"
"MaxTransDiff" 1)

;---------- Save BND and CMD and rescale to nm ------;

(sde:assign-material-and-region-names (get-body-list) )

(sdeio:save-tdr-bnd (get-body-list) "n@node@_nm.tdr")

(sdedr:write-scaled-cmd-file "n@node@_msh.cmd" nm)

(define sde:scale-tdr-bnd

 (lambda (tdrin sf tdrout)

  (sde:clear)

  (sdegeo:set-default-boolean "XX")


  (sdeio:read-tdr-bnd tdrin)

  (entity:scale (get-body-list) sf)

  (sdeio:save-tdr-bnd (get-body-list) tdrout)

 )

)
(sde:scale-tdr-bnd "n@node@_nm.tdr" nm "n@node@_bnd.tdr")

;-------------- END --------------------;
```

See Figs. 3.54 and 3.55

Example 3.6 InGaAs nFinFET with L_g = 100 nm and HfO$_2$/Al$_2$O$_3$ as gate insulator

III–V semiconductor materials have been promised as a high mobility solution for the sub-7-nm technology node. Table 3.4 shows mobility of important semiconductor materials for mass production. **The electron mobility of InAs is 30,000 cm^2/Vs and InGaAs is 11,200 cm^2/Vs.**

In order for the promise of IIIV to be finally realized as a viable option for CMOS co-integration or high-density heterogeneous devices, these materials must be integrated on 300 mm or larger Si substrates in a fully VLSI compatible flow. The InGaAs has highest electron mobility; thus, it is suitable for next-generation nFinFET to replace Si nFinFET. According to Fig. 3.56, there has only **8% lattice mismatch between (InP, In$_{0.53}$Ga$_{0.47}$As) and Si.** Therefore, **In$_{0.53}$Ga$_{0.47}$As** can integrate of Si substrate by using InP buffer layer.

In Example 3.6, we study the In$_{1-x}$Ga$_x$As FinFET on Si substrate with InP buffer layer, as shown in Fig. 3.57. In$_{1-x}$Ga$_x$As FinFET applies different molecular fractions by using high-k material HfO$_2$/Al$_2$O$_3$ (bottom) of 3 nm/3 nm, based on 2014 IMEC's 2014 VLSI technical paper [6]. The In$_{1-x}$Ga$_x$As molecular fraction (xF) changes from 0 to 1, step 0.2, and **0.47** which are shown in Fig. 3.58.

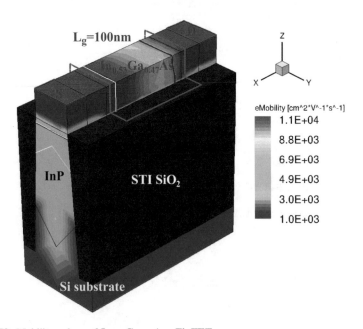

Fig. 3.59 Mobility values of **In$_{0.53}$Ga$_{0.47}$As** nFinFET

The default Molefraction.txt file has the following content:

2. SWB tool SDEVICE → dessis_des.cmd

First, in SEVICE tool must include parameter files, SiliconGermanium and HfO2. The dessis_des.cmd is identical to example 3.1, only add following text

Physics(material="SiliconGermanium"){

   MoleFraction(xFraction=@xF@)

}

In(1-x)Ga(x)As

InGaAs (x=0) = InAs

InGaAs (x=1) = GaAs

For $x = 0.47$, $In(1-x)Ga(x)As$ is $In_{0.53}Ga_{0.47}As$. In this example, we set a variable @xF@ as a molecular fraction, as identical parameter as above x.

Figure 3.59 shows the mobility of $In_{0.53}Ga_{0.47}As$ nFinFET. The mobility value of $In_{0.53}Ga_{0.47}As$ whole channel is larger than 8000 cm²/Vs.

The 1D energy band diagram of $In_{0.53}Ga_{0.47}As$ nFinFET for X and Y directions is shown in Fig. 3.60a, b, respectively.

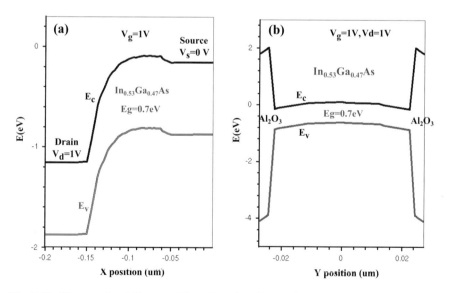

Fig. 3.60 1D energy band diagram of $In_{0.53}Ga_{0.47}As$ nFinFET for **a** X and **b** Y direction

The 1D energy band diagram of $In_{0.53}Ga_{0.47}As$ nFinFET for Z direction is shown in Fig. 3.61a. The I_d–V_g transfer curve of $In_{0.53}Ga_{0.47}As$ nFinFET is shown in Fig. 3.61b. The L_g = 100 nm $In_{0.53}Ga_{0.47}As$ nFinFET on-state current (I_{sat}) is 468 μA much larger Si L_g = 15 nm FinFET of 24.2 μA of Example 3.1. The simulation results are consistent with the intrinsic semiconductor material properties, especially in mobility.

In summary, the standard example of TCAD simulation of 3D FinFET is provided in this chapter, which includes nFinFET and pFinFET with L_g = 15 nm. It can serve as important reference for the research and development of 3D FinFET. It suggests its potential for contributing to future development of 3D FinFET of sub-10-nm semiconductor technology node. In addition, we simulation the L_g = 10 nm nFinFET with different F_h and V_{dd} = 0.7 V. For F_h = 35 nm, the I_{off} is around 1E−9 A, and I_{sat} is around 4.5E−5 A. It would not meet requirements ITRS 2.0 electrical specification. In final two Examples 3.5 and 3.6, we study high mobility $Si_{1-x}Ge_x$ and $In_{0.53}Ga_{0.47}As$ FinFET. In another approach, we may need more electrical control gate-all-around (GAA) nanowire FET to achieve ITRS requirements. For the reason, we will discuss the L_g = 10 nm GAA NWFET simulation in Chap. 5.

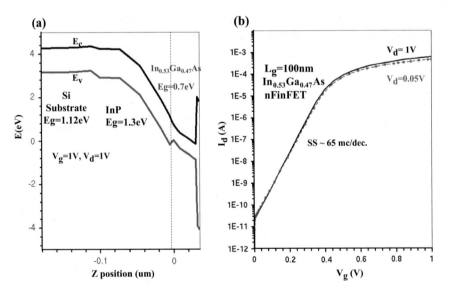

Fig. 3.61 **a** 1D energy band diagram of $In_{0.53}Ga_{0.47}As$ nFinFET for Z direction. **b** I_d–V_g of $In_{0.53}Ga_{0.47}As$ nFinFET

References

1. C.C. Hu, in *Modern Semiconductor Devices for Integrated Circuits* (PEARSON, 2010)
2. C.H. Jan, U. Bhattacharya, R. Brain, S.-J. Choi, G. Curello, G. Gupta, W. Hafez, M. Jang, M. Kang, K. Komeyli, T. Leo, N. Nidhi, L. Pan, J. Park, K. Phoa, A. Rahman, C. Staus, H. Tashiro, C. Tsai, P. Vandervoorn, L. Yang, J.-Y. Yeh, P. Bai, A *22 nm* SoC platform technology featuring 3-D tri-gate and high-k/metal gate, optimized for ultra low power, high performance and high density SoC applications. Tech. Digest IEDM 3.1.1 (2012)
3. S. Sinha, B. Cline, G. Yeric, V. Chandra, Y. Cao, Design benchmarking to 7 nm with FinFET predictive technology models. *ISLPED* **15** (2012)
4. M. Bohr, *Opening New Horizons: 14 nm Process Technology* (Intel IDF, 2014)
5. J.P. Colinge, in *FinFETs and Other Multi-Gate Transistors* (Springer, 2007)
6. N. Waldron, C. Merckling, W. Guo, P. Ong, L. Teugels, S. Ansar, D. Tsvetanova, F. Sebaai, D.H. van Dorp, A. Milenin, D. Lin, L. Nyns, J. Mitard, A. Pourghaderi, B. Douhard, O. Richard, H. Bender, G. Boccardi, M. Caymax, M. Heyns, W. Vandervorst, K. Barla, N. Collaert, A.V.Y. Thean, An InGaAs/InP quantum well finfet using the replacement fin process integrated in an RMG flow on 300 mm Si substrates. VLSI Tech. Symp. **1** (2014)

Chapter 4
Inverter and SRAM of FinFET with L_g = 15 nm Simulation

The fundamental of transistors of the circuits can be simulated by using L_g = 15 nm FinFET CMOS Inverter and static random-access memory (SRAM). They will consume small amounts of power, and they are equipped with important characteristics of regenerating or cleaning up digital signals. Such FinFET CMOS Inverter basic characteristics will discuss in details in Sect. 4.1, and the speed of Inverter will discuss in Sect. 4.2. SRAM only requires the same transistors and fabrication processes of the basic CMOS technology. It is therefore the easiest to integrate or embed into COMS circuits. The SRAM simulation based on L_g = 15 nm FinFET will be discussed in Sect. 4.6 [1–3].

4.1 Voltage Transfer Curve of Inverter

Consider the CMOS Inverter shown in Fig. 4.1a. The IV curve of nFET is as shown on the right half of 4.1b. Assume that the pFET has identical (symmetric) IV as plotted on the left half of the figure. From (a), the V_{ds} of the pFET and nFET are related to V_{out} by $V_{dsN} = V_{out}$ and $V_{dsP} = V_{out} - 2$ V. Therefore, the two halves of (b) can be replotted in (c) using V_{out} as the common variable. For example, at V_{out} = 2 V in (c), V_{dsN} = 2 V and V_{dsP} = 0 V.

The two V_{in} = 0 V curves in Fig. 4.1c intersect at V_{out} = 2 V. This means V_{out} = 2 V when V_{in} = 0. This point is recorded in Fig. 4.2. The two V_{in} = 0.5 V curves intersect at around V_{out} = 1.9 V. The two V_{in} = 1 V curves intersect at V_{out} = 1 V. All the V_{in}/V_{out} pairs are represented by the curve in Fig. 4.2, which is the voltage transfer characteristic of Inverter or **voltage transfer curve (VTC)**. The VTC provides digital circuit with important noise margin. V_{in} can be anywhere from 0 V to V_{th} of nFET while resulting in the ideal V_{out} = V_{dd}. Similarly, V_{in} can be anywhere between 2 and 2 V plus V_{th} of pFET while resulting in the ideal V_{out} = 0 V.

© Springer Nature Singapore Pte Ltd. 2018
185
Y.-C. Wu and Y.-R. Jhan, *3D TCAD Simulation for CMOS Nanoeletronic Devices*,
DOI 10.1007/978-981-10-3066-6_4

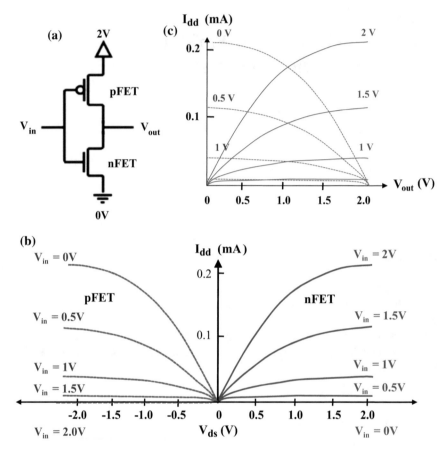

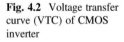

Fig. 4.1 a CMOS inverter, **b** IV characteristics of nFET and pFET, and **c** $V_{out} = V_{dsN} = 2\ V + V_{dsP}$ according to (**a**)

Fig. 4.2 Voltage transfer curve (VTC) of CMOS inverter

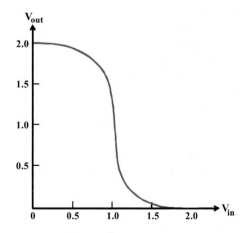

Therefore, perfect "0" and "1" outputs can be produced by somewhat corrupted inputs. This regenerative property allows complex logic circuits to function properly in the face of inductive and capacitive noises and IR drops in the signal lines. A VTC with a narrow and steep middle region will maximize the noise tolerance. Device characteristics which can be used to generate ideal VTC include greater g_m, low leakage current in the off-state, and small $\Delta I_{ds}/\Delta V_{ds}$ in saturation region. The two device characteristics will be further discussed in the next section.

During the operation of an ideal circuit, the transition region of VTC should be located at or near $V_{in} = V_{dd}/2$. For achieving the symmetry, the IV curves of nFET and pFET in Fig. 4.1b must be matched while being folded. This is achieved by choosing the width (W) of transistors, W value of pFET greater than the W value of nFET. Generally speaking, the ratio of **W_p/W_N is around 3** in order to compensate the ratio of **$\mu_{ps}/\mu_{ns} = 3$** in $L_g = 15$ nm FinFET.

4.2 Speed of CMOS Inverter—Importance of I_{on}

The propagation delay is the **delay time τ_d** required by the signal to propagate from one logic gate to the next identical logic gate as shown in Fig. 4.3.

τ_d is the average of falling delay (rising V_1 pulling down the output, V_2) and rising delay (falling V_2 pulling up the output, V_3). The propagation delay of the Inverter is expressed below:

Fig. 4.3 a A CMOS inverter chain, and the transistor with circle symbol at gate G indicates pFET, **b** propagation delay of τ_d

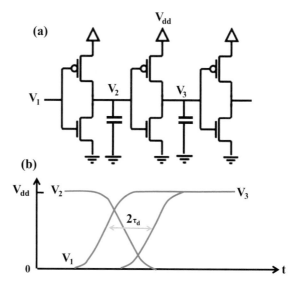

$$\tau_d \approx \frac{CV_{dd}}{4}\left(\frac{1}{I_{onN}} + \frac{1}{I_{onP}}\right) \tag{4.1}$$

I_{onN} is taken at $V_{gs} = V_{dd}$ and I_{onP} taken at $V_{gs} = -V_{dd}$. They are called the **on-state current** of the nFET and the pFET.

$$I_{on} \equiv I_{dsat}\big|_{\max|v_{gs}|} \tag{4.2}$$

There is a simple explanation of Eq. (4.1)

$$\tau_D = \frac{1}{2}(\tau_F\,\text{pull} - \text{down delay} + \tau_R\,\text{pull} - \text{up delay}) \tag{4.3}$$

$$\text{pull} - \text{down delay } \tau_F \approx \frac{CV_{dd}}{2I_{onN}}$$
$$\text{pull} - \text{up delay } \tau_R \approx \frac{CV_{dd}}{2I_{onP}} \tag{4.4}$$

The total delay refers to the time required for a conducting transistor to provide an I_{on} current to change the output by $V_{dd}/2$ (not V_{dd}) in Fig. 4.3b. The charge drained from C by the FET during the delay is $CV_{dd}/2$. Therefore, the delay is $\tau_d = Q/I = CV_{dd}/2I_{on}$. We can regard τ_d as RC delay, and $V_{dd}/2I_{on}$ as transistor switching resistance of the transistor. For maximizing the circuit operation speed, I_{on} must be maximized, and the electric properties of pFET and nFET must be in perfect symmetry.

4.3   CMOS I_d–V_g Matching Diagram for High-Performance Transistors

As CMOS IC technology scales down to the 14-nm node, the FinFET transistor technology has proven its superior capability to enable very aggressive and following Moore's law. The 14-nm technology with a wide range of system-on-chip (SoC) products, including tablets, smart phones, ASIC, embedded, Internet-of-Things, baseband, and RF products [3].

The high performance (HP), standard performance (SP), and ultra-low power (ULP) logic transistors are fully compatible with different SoC applications show that all transistors exhibit well-behaved electrical characteristics including I_d–V_d, I_d–V_g, DIBL, and sub-threshold slope (SS).

The nFinET and pFinET inside the Inverter must allow I_d–V_g to be perfectly matched (folded symmetry).

$$|V_{TN}| = |V_{TP}| \text{ and } I_{on|p} = I_{on|n} \qquad (4.5)$$

Assuming fin width F_w = **5 nm** and considering the conduction in nFinET is based on electrons, the electron mobility is roughly 2–3 times to the hole mobility in the transistor, the F_w of pFinET should be adjusted to math Eq. (4.5) of that in nFinFET as shown in Table 4.1, meaning F_w = **16 nm** of pFinFET with $|V_{TN}| = |V_{TP}| = $ **0.30 V** as shown in Fig. 4.4. The ratio of W_p/W_N **is around 3** in order to compensate the ratio of μ_{ps}/μ_{ns} = **3** in this L_g = 15 nm 3D FinFET Inverter example, the width ratio is W_p/W_N = **16 nm/5 nm** for ideal CMOSFET matching as shown in Fig. 4.4.

4.4 [Example 4.1] Inverter of 3D FinFET with L_g = 15 nm

The nFinFET and pFinFET program code file is identical to Example 3.1 (or Table 4.1, 4.1a) and Example 3.3 (or Table 4.1, 4.1c F_w/F_h = 16 nm/5 nm) of Chap. 3, respectively. Example 4.1 is hereby combined nFinFET and pFinFET to form the entire SDE codes of Inverter.

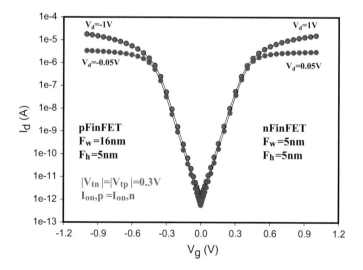

Fig. 4.4 I_d–V_g characteristic in which the electric properties of nFinFET (Example 3.1) and pFinFET (Example 3.3) must be perfectly matched for the inverter to be equipped with excellent properties

Table 4.1 Electric properties matching parameters of nFinFET and pFinFET extracted from Fig. 4.4

Example	$V_d = 1$ V	F_w/F_h (nm)	WF (V)	V_t (V)	SS (mV/dec)	I_{sat} (A)
4.1a	nFinFET	5/5	4.48	0.3020	65.8	2.685E−5
4.1b	pFinFET	5/5	4.82	−0.3038	65.2	1.853E−5
4.1c	pFinFET	16/5	4.80	−0.2939	68.3	2.69e−05

The width ratio is W_p/W_N = **16 nm/5 nm–3** for ideal CMOSFET matching

Fig. 4.5 Required simulation tools are shown in the workbench of inverter simulation on SWB

The following three main tools and their code files are based on Synopsys Sentaurus TCAD 2014 version (tool → *.cmd)
   SDE → devise_dvs.cmd, SDEVICE → dessis_des.cmd, and INSPECT → inspect_inc.cmd.
   Figure 4.5 is simulation tools of SWB for Inverter simulation.
   Figure 4.7 is the simulated input and output electrical properties of Inverter based on 3D FinFET with L_g = 15 nm.

1. **SDE – devise_dvs.cmd**

   The SDE CODE of Examples 3.1 and 3.3 **is substituted into nFET and pFET, respectively. The first two nodes are for nFET, and the following two nodes are for pFET.**
   Therefore, here we only need to focus on the following SDEVICE program codes. Here, we use SDEVICE default library example.

2. **SDVICE – dessis_des.cmd**

```
* ------------------- Ex 4.1 dessis_des.cmd of SDEVICE -------------------*
NMOS {
Electrode{
{ Name="S" Voltage=0.0 }
{ Name="D" Voltage=0.0 }
{ Name="G" Voltage=0.0 Workfunction= @WK@}
}
File{
Grid      = "@tdr|-2@"
Plot      = "@tdrdat@"
Current  = "@plot@"
*Output  = "@log@"
}
Physics{
Mobility(DopingDepHighFieldSaturationEnormal )
EffectiveIntrinsicDensity(OldSlotboom )
Recombination( SRH(DopingDep) )
}}
SDEVICEPMOS{
Electrode{
{ Name="S" Voltage=0.0 }
{ Name="D" Voltage=0.0 }
{ Name="G" Voltage=0.0 Workfunction= @WK1@ }
}
File{
Grid = "@tdr@"
Plot = "@tdrdat@"
Current = "@plot@"
*Output = "@log@"
}
Physics{
Mobility(DopingDepHighFieldSaturationEnormal )
EffectiveIntrinsicDensity(OldSlotboom )
Recombination( SRH(DopingDep) )
```

```
}}
File{
Output = "@log@"
}
Plot{
*---------------------------Density and Currents, etc
eDensityhDensity
TotalCurrent/Vector eCurrent/Vector hCurrent/Vector
eMobilityhMobility
eVelocityhVelocity
eQuasiFermihQuasiFermi
*---------------------------Fields and charges
ElectricField/Vector Potential SpaceCharge
*---------------------------Doping Profiles
Doping DonorConcentrationAcceptorConcentration
*---------------------------Generation/Recombination
SRH Auger
* AvalancheGenerationeAvalancheGenerationhAvalancheGeneration
*---------------------------Driving forces
eGradQuasiFermi/VectorhGradQuasiFermi/Vector
eEparallelhEparalllel
*---------------------------Band structure/Composition
BandGap
BandGapNarrowing
Affinity
ConductionBandValenceBand
}
Math (Region="Channel") {Nonlocal(-Transparent)
}
Math{
Extrapolate
Derivatives
* Avalderivatives
RelErrControl
```

```
Digits=5
ErRef(electron)=1.e10
ErRef(hole)=1.e10
Notdamped=50
Iterations=20
*Newdiscretization
Directcurrent
Method=ParDiSo
Parallel= 2
*-VoronoiFaceBoxMethod
NaturalBoxMethod
}
System{
Vsource_pset VVDD (vdd 0) { dc = 0 }
Vsource_pset VGND (gnd 0)   { dc=0 }
Vsource_pset VVIN (in 0)     { pulse = ( 0 @Vdd@ 0.3e-10 0.02e-10
0.02e-100.3e-10 3000 )}        (1)   (2)      (3)      (4)
(5)        (6)      (7)


NMOS nmos1 (    "D"=out "G"=in "S"=gnd   )
PMOS pmos1 (    "D"=out "G"=in "S"=vdd   )
Plot "n@node@_sys_des.plt" (time() v(in) v(out)i(nmos1,out) i(pmos1,out))
}
* ------------------------------- Remark -------------------------------*
```

Note: The aforementioned V_{in} input square wave diagram is as shown on the left of Fig. 4.6, and the Inverter circuit diagram is on the right.

(1). (2). … (7) are the program codes for definition of input square wave.

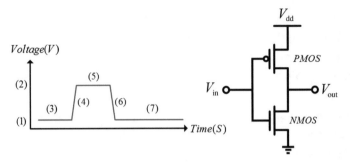

Fig. 4.6 V_{in} input square wave diagram and inverter circuit diagram. *1* lowest voltage, *2* V_{dd}, *3* delay time, *4* rising time, *5* falling time, *6* duration time, and *7* period time

Solve{

Coupled(Iterations=150){ Poisson }

Coupled{ Poisson}

Coupled{ Poisson Electron Hole Contact Circuit }

Quasistationary(

InitialStep=1e-3 Increment=1.35

MinStep=1e-8 MaxStep=0.05

Goal{ Parameter=VVDD.dc Voltage= @Vdd@ })

{Coupled{nmos1.poisson nmos1.electron nmos1.hole

nmos1. nmos1.contact   pmos1.poisson   pmos1.electron pmos1.hole

pmos1.pmos1.contact circuit }}

NewCurrentfile = "TR_"

Transient(

InitialTime=0

FinalTime=0.8e-10

InitialStep=1e-12

MaxStep=1e-11

Minstep=1.e-18

Increment=1.1)

{ Coupled{nmos1.poisson   nmos1.electron   nmos1.hole nmos1.

nmos1.contactpmos1.poissonpmos1.electron pmos1.hole pmos1. pmos1.contact

   circuit }}}

* ------------------------------------ END --*

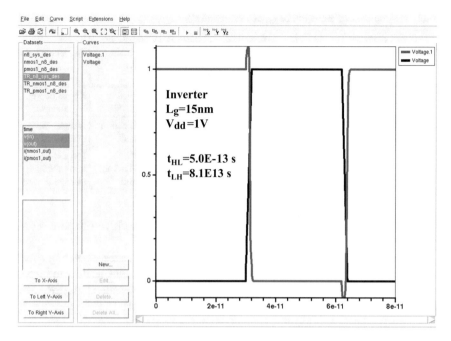

Fig. 4.7 Inverter voltage–time curve of L_g = 15 nm 3D FinFET

Figure 4.7 is the simulated input and output electric properties of Inverter of 3D FinFET with L_g = 15 nm.

4.5 TCAD Simulation of Static Random-Access Memory (SRAM)

There are three different types of semiconductor memory: static random-access memory (SRAM), dynamic random-access memory (DRAM), and the nonvolatile memory nowadays (or Flash memory). "Nonvolatile memory" means that the data will not be lost when the power of memory is turned off. These three kinds of memories can coexist because each of them is equipped with unique advantages and limits. The differences among them are summarized in Table 4.2.

SRAM is composed of the most fundamental transistors without the need for additional complicated process, thus making it easy to be integrated or embedded in the logic circuit. The difference between SRAM and DRAM is: SRAM is based on a more complicated structure with less capacity per unit area and fast access speed; DRAM is based on a rather simple structure with large capacity per unit area, but it has slower access speed. Although DRAM has simpler structure than SRAM, the

Table 4.2 Differences among three types of memories

	Is power required during data saving?	Unit size and cost/bit	Overwrite cycle	Speed of writing a byte	Compatible with basic CMOS	Major applicable field
SRAM	Yes	High	Unlimited	Fast	Completely compatible	Embedded in logic chip
DRAM	Yes	High	Unlimited	Fast	Modification required	Independent or embedded chip
Flash memory	No	Lowest	Limited	Slow	Major modification required	Independent nonvolatile storage device

stored charges can gradually disappear as time goes by, thus it will require certain refreshing to keep the data stored in the capacitors.

Among all memories in Table 4.2, SRAM provides the fastest operating speed. However, it will take six transistors to store one bit of data, thus leading to highest cost per bit. When the processor speed is an important consideration, SRAM is often used as the cache memory embedded in the processor.

4.6   SRAM Operation

In CMOS VLSI designs, the most commonly used SRAM storage element is bistable latch consisting of two cross-coupled CMOS Inverters shown in Fig. 4.8. It can be built using a standard CMOS logic fabrication process. Inverter 1 consists of nFinFET Q_1 and pFinFET Q_3 while Inverter 2 consists of nFinFET Q_2 and pFinFET Q_4. The two stable states can be readily recognized by plotting the transfer curves of the two Inverters back to back, as illustrated in Fig. 4.9, often referred to as the "butterfly curve" plot of a pair of cross-coupled Inverters.

In Fig. 4.8, one of the Inverters has its input at high and output at low, while the other Inverter has its input at low and output at high. The first Inverter, with its output at low, keeps the second Inverter in the state described above, and vice versa. Thus, a CMOS SRAM storage element has two stable states: one at intersection A of the two Inverter transfer curves in Fig. 4.9 with $V_1 = V_{in2} = V_{dd}$, and the other at the intersection B with $V_2 = V_{in1} = V_{dd}$. The two stable states can be interpreted as logical "0" and "1". Here, we designate **logical "1" as $V_1 = 0$ and $V_2 = V_{dd}$**, i.e., point B, and **logical "0" as $V_1 = V_{dd}$ and $V_2 = 0$**, i.e., point A. A bistable latch will remain in one of its two stable states until it is forced by an external signal to flip to the other stable state.

The most commonly used SRAM cell is a six-transistor cell consisting of two cross-coupled CMOS Inverters and two access transistors. The circuit schematic for CMOS SRAM cell is shown in Fig. 4.8. The cross-coupled Inverters are connected to two bitlines, BLT (bitline true) and BLC (bitline complement), through

Fig. 4.8 a Circuit diagram of
SRAM unit **b** SRAM layout,
where the 14 *small rectangles*
are connecting points, the *four
horizontal rectangles* are
gates G, and the width of two
pFinFETs channels are
greater than four nFinFETs

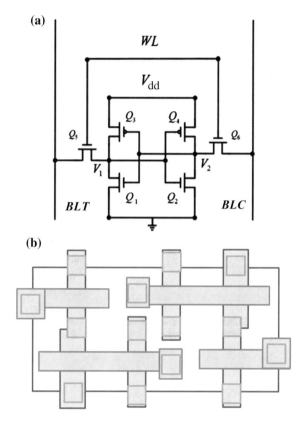

n-channel access transistors Q_5 and Q_6. The access transistors are controlled by the word line (WL) voltage. In the standby mode, WL is kept low ($V_{WL} = 0$ V), thus turning off the access transistors and isolating the bitlines from the cross-coupled Inverters pair.

The two switch transistors, Q_5 and Q_6, are connecting the output of Inverter to the bitline. For reading the saved data (by determining the state of Inverter), the WL of the selected unit will be raised to high potential to turn on the access transistor. A sensitive sensing amplifier circuit will compared the voltage differences of BLT and BLC to determine the saved state. In order to write the low state "0" into the unit on the left, BLT will be set at low potential and BLC will be set at high potential.

The example of SRAM Read operation: SRAM operation for reading "0" is as shown in Fig. 4.10. **Where** $V_1 = V_{dd} = 1$ **V, and** $V_2 = 0$ **V.**

The reading and writing of SRAM will be executed via conduction of word line (WL) ("Logic 1") (conduction between node Q_5 and Q_6). Here, we assume the data saved in the V_2 of SRAM unit is "Logic 0".

The reading operation process is as shown below:

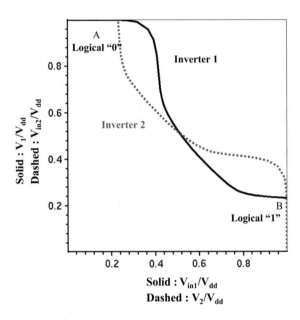

Fig. 4.9 Butterfly plot for two cross-coupled CMOS Inverters. The transfer curve of inverter 1 (*solid*) is plotted as V_1 versus V_{in1}, and inverter 2 (*dashed*) as V_{in2} versus V_2

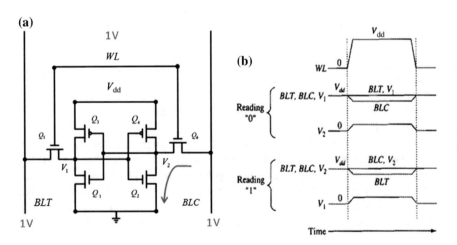

Fig. 4.10 a Relationship between voltage and current during CMOS SRAM operation for reading "0", **b** the electric potential analysis of Node V_1 and Node V_2 during CMOS SRAM operation for reading "0"

(1) Before reading: the node voltage V_1 is logic "1" (V_{dd}) and V_2 is logic "0"
     (0 V).
(2) The parasitic capacitors of the two lines of BLT and BLC are pre-charged to
     logic 1 such that the node voltage is $V_{BLT} = V_{BLC} = V_{dd}$, than selected WL
     turn on.
(3) Q_5 transistor is not conducting because the voltage on two terminals
     $V_{BLT} = V_1 = V_{dd}$.
(4) In Q_6 transistor, the capacitor C_{V2} occurs charge sharing because the voltage
     on two terminals $V_{BLC} = V_{dd}$ with $V_2 = 0$ V.
(5) As for the bitline (BL), V_{BLC} will be less than V_{BLT} due to charge sharing and
     capacitive voltage division effects. When the difference between them is as
     high as ΔV (around 0.1–0.2 V), it will trigger the sense amplifier to amplify
     the signal difference. After it is converted into the output, it will be sent to the
     data buffer to complete the operation of reading "0". Even though there can be
     interference during reading, the positive feedback of latch can restore V_2 to
     0 V, which is logic 0.
(6) After the operation is completed, WL will turn off to 0 V such that this SRAM
     will be disconnected from BLs, thus ending the entire reading operation.

**SRAM Write operation is as described below: it is the example of SRAM
operation for writing "1" (from the original state of "0" to "1") is as shown in
Fig. 4.11.**

The reading and writing of SRAM will be executed via conduction of word line
(WL) ("Logic 1") (conduction between node Q_5 and Q_6). Here, we assume the data
saved in the V_2 of SRAM unit is "Logic 0", and now "Logic 1" must be written.

The writing operation process is as shown below:

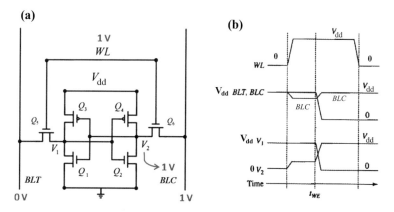

Fig. 4.11 a Relationship between voltage and current of CMOS SRAM during the operation of
writing "1" **b** the electric potential analysis of Node V_1 and Node V_2 during CMOS SRAM
operation for writing "1"

(1) Before writing: the node voltage V_1 is logic "1" (V_{dd}) and V_2 is logic "0" (0 V).

(2) The parasitic capacitor of node BLT is reduced to logic "0", and the parasite capacitor of node BLC is pre-charged to logic "1", such that node voltage V_{BLT} = 0 V; V_{BLC} = V_{dd}, than selected WL turn on.

(3) As for transistors Q_5 and Q_3, there is current flowing through Q_3 into Q_5.

(4) As for transistors Q_6 and Q_2, capacitor C_{V2} can be rapidly charged via Q_6.

(5) As for the bitlines (BLT, BLC), the voltage on BLT forces V_1 to "0", while the voltage on BLC forces V_2 to V_{dd}, thus writing a logic "1" to the cell.

(6) After the operation is completed, WL will turn off to 0 V such that this SRAM will be disconnected from BLs, thus ending the entire writing operation.

4.7   [Example 4.2] Simulation of SRAM of 3D FinFET with L_g = 15 nm

The identical program codes of nFinFET of Example 3.1 (or Table 4.1, 4.1a) and pFinFET of Example 3.3 (or Table 4.1, 4.1c F_w/F_h = 16 nm/5 nm) to form the SRAM as shown below. SRAM is composed of 6 FinFETs (4 nFinFETs and 2 pFinFETs) as shown in Fig. 4.8a.

The following three main tools and their code files are based on Synopsys Sentaurus TCAD 2014 version (tool → *.cmd)

SDE → NPNPNN-FET six devise_dvs.cmd, SDEVICE → dessis_des.cmd, and INSPECT → inspect_inc.cmd.

1. SDE – devise_dvs.cmd

As shown in Fig. 4.12 workbench, in this example with a sequence of (nFET-pFET: Inverter1), (nFET-pFET: Inverter2), and nFET-nFET (access transistors) SRAM, we only explain the following program codes of SDEVICE. Here, we use SDEVICE default library example.

2. SDEVISE – dessis_des.cmd

Fig. 4.12 Required simulation tools are shown in the workbench for SRAM simulation

```
* ---------------------- Ex 4.2 dessis_des.cmd ----------------------*
NMOS1 {
Electrode{
{ Name="S"Voltage=0.0 }
{ Name="D"Voltage=0.0 }
{ Name="G" Voltage=0.0 Workfunction= @WKN@}
*{ Name="B" Voltage=0.0 }
}
File{
Grid = "@tdr|-10@"
Plot = "@tdrdat@"
Current = "@plot@"
Output= "@log@"
}
Physics{
Mobility(DopingDepHighFieldSaturationEnormal )
EffectiveIntrinsicDensity(OldSlotboom )
Recombination( SRH(DopingDep) )
}}
Device PMOS1{
Electrode{
{ Name="S" Voltage=0.0 }
{ Name="D"    Voltage=0.0 }
{ Name="G"    Voltage=0.0 Workfunction= @WKP@ }
* { Name="B" Voltage=0.0 }
}
File{
Grid= "@tdr|-8@"
Plot = "@tdrdat@"
Current = "@plot@"
*Output = "@log@"
}
Physics{
Mobility(DopingDepHighFieldSaturationEnormal )
```

```
EffectiveIntrinsicDensity(oldSlotboom )
Recombination( SRH(DopingDep) )
}}
Device NMOS2 {
Electrode{
{ Name="S" Voltage=0.0 }
{ Name="D" Voltage=0.0 }
{ Name="G" Voltage=0.0 Workfunction= @WKN@}
*{ Name="B" Voltage=0.0 }
}
File{
Grid = "@tdr|-6@"
Plot = "@tdrdat@"
Current = "@plot@"
*Output= "@log@"
}
Physics{
Mobility(DopingDepHighFieldSaturationEnormal )
EffectiveIntrinsicDensity(OldSlotboom )
Recombination( SRH(DopingDep) )
}}
Device PMOS2{
Electrode{
{ Name="S" Voltage=0.0 }
{ Name="D" Voltage=0.0 }
{ Name="G" Voltage=0.0 Workfunction= @WKP@ }
*{ Name="B" Voltage=0.0 }
}
File{
Grid = "@tdr|-4@"
Plot = "@tdrdat@"
Current = "@plot@"
*Output = "@log@"
}
```

```
Physics{
Mobility(DopingDepHighFieldSaturationEnormal )
EffectiveIntrinsicDensity(oldSlotboom )
Recombination( SRH(DopingDep) )
}}
Device NMOS3 {
Electrode{
{ Name="S" Voltage=0.0 }
{ Name="D" Voltage=0.0 }
{ Name="G" Voltage=0.0 Workfunction= @WKN@}
* { Name="B" Voltage=0.0 }
}
File{
Grid = "@tdr|-2@"
Plot = "@tdrdat@"
Current = "@plot@"
*Output= "@log@"
}
Physics{
Mobility(DopingDepHighFieldSaturationEnormal )
EffectiveIntrinsicDensity(OldSlotboom )
Recombination( SRH(DopingDep) )
}}
Device NMOS4{
Electrode{
{ Name="S" Voltage=0.0 }
{ Name="D" Voltage=0.0 }
{ Name="G" Voltage=0.0 Workfunction= @WKN@ }
* { Name="B" Voltage=0.0 }
}
File{
Grid = "@tdr@"
Plot = "@tdrdat@"
Current = "@plot@"
```

```
*Output = "@log@"
}
Physics{
Mobility(DopingDepHighFieldSaturationEnormal )
EffectiveIntrinsicDensity(oldSlotboom )
Recombination( SRH(DopingDep) )
}}
File{
Output = "@log@"
}
Plot{
*--Density and Currents, etc
eDensityhDensity
TotalCurrent/Vector eCurrent/Vector hCurrent/Vector
eMobilityhMobility
eVelocityhVelocity
eQuasiFermihQuasiFermi
*--Fields and charges
ElectricField/Vector Potential SpaceCharge
*--Doping Profiles
Doping DonorConcentrationAcceptorConcentration
*--Generation/Recombination
SRH Auger
* AvalancheGenerationeAvalancheGenerationhAvalancheGeneration
*--Driving forces
eGradQuasiFermi/VectorhGradQuasiFermi/Vector
eEparallelhEparalllel
*--Band structure/Composition
BandGap
BandGapNarrowing
Affinity
ConductionBandValenceBand }

Math{
```

```
Extrapolate
Derivatives
* Avalderivatives
RelErrControl
Digits=5
ErRef(electron)=1.e10
ErRef(hole)=1.e10
Notdamped=50
Iterations=20
*Newdiscretization
Directcurrent
Method=ParDiSo
Parallel= 2
*-VoronoiFaceBoxMethod
NaturalBoxMethod }

System{
Vsource_psetvdd (dd 0) { dc = 0.0 }
Vsource_psetvwl (T 0) { dc = 0.0 }
Vsource_psetvb (L 0) { dc = 0.0 }
Vsource_psetvbl (R 0) { dc = 0.0 }
Vsource_pset vin (VinL 0) { dc = 0.0 }
NMOS1 nmos1( "S"=0    "D"=VinR "G"=VinL   )
NMOS2 nmos2( "S"=0    "D"=VinL "G"=VinR   )
PMOS1 pmos1( "S"=dd "D"=VinR "G"=VinL   )
PMOS2 pmos2( "S"=dd "D"=VinL "G"=VinR   )
NMOS3 nmos3( "S"=VinR    "D"=L "G"=T   )
NMOS4 nmos4( "S"=R    "D"=VinL "G"=T   )
Plot "n@node@_sys_des.plt" (time() v(VinL) v(dd) v(T) v(L) v(R) v(VinR))}
Solve{
NewCurrentFile="init"
Coupled(Iterations=250){ Poisson   }
Coupled{ Poisson Electron Hole Contact Circuit   }
Quasistationary(
```

```
InitialStep=1e-3 Increment=1.35
MinStep=1e-12 MaxStep=0.05
Goal{ Parameter=vdd.dc Voltage=@Vdd@ }
Goal{ Parameter=vwl.dc Voltage= @Vdd@ }
Goal{ Parameter=vb.dc Voltage= @Vdd@ }
Goal{ Parameter=vbl.dc Voltage= @Vdd@}
){ Coupled{ nmos1.poisson nmos1.electron nmos1. nmos1.contact
nmos2.poisson nmos2.electron nmos2. nmos2.contact
nmos3.poisson nmos3.electron nmos3. nmos3.contact
nmos4.poisson nmos4.electron nmos4. nmos4.contact
pmos1.poisson pmos1.hole pmos1. pmos1.contact
pmos2.poisson pmos2.hole pmos2. pmos2.contact
circuit }
}
Quasistationary(
InitialStep=1e-3 Increment=1.35
MinStep=1e-12 MaxStep=0.05
Goal{ Parameter=vdd.dc Voltage=@Vdd@ }
){ Coupled{ nmos1.poisson nmos1.electron nmos1. nmos1.contact
nmos2.poisson nmos2.electron nmos2. nmos2.contact
nmos3.poisson nmos3.electron nmos3. nmos3.contact
nmos4.poisson nmos4.electron nmos4. nmos4.contact
pmos1.poisson pmos1.hole pmos1. pmos1.contact
pmos2.poisson pmos2.hole pmos2. pmos2.contact
circuit }
}
Quasistationary(
InitialStep=1e-3 Increment=1.35
MinStep=1e-12 MaxStep=0.05
Goal{ Parameter=vwl.dc Voltage= @Vdd@ }
){ Coupled{ nmos1.poisson nmos1.electron nmos1. nmos1.contact
nmos2.poisson nmos2.electron nmos2. nmos2.contact
nmos3.poisson nmos3.electron nmos3. nmos3.contact
nmos4.poisson nmos4.electron nmos4. nmos4.contact
```

pmos1.poisson pmos1.hole pmos1. pmos1.contact

pmos2.poisson pmos2.hole pmos2. pmos2.contact

circuit }

}

Quasistationary(

InitialStep=1e-3 Increment=1.35

MinStep=1e-12 MaxStep=0.05

Goal{ Parameter=vb.dc Voltage= @Vdd@ }

){ Coupled{ nmos1.poisson nmos1.electron nmos1. nmos1.contact

nmos2.poisson nmos2.electron nmos2. nmos2.contact

nmos3.poisson nmos3.electron nmos3. nmos3.contact

nmos4.poisson nmos4.electron nmos4. nmos4.contact

pmos1.poisson pmos1.hole pmos1. pmos1.contact

pmos2.poisson pmos2.hole pmos2. pmos2.contact

circuit }

}

Quasistationary(

InitialStep=1e-3 Increment=1.35

MinStep=1e-12 MaxStep=0.05

Goal{ Parameter=vbl.dc Voltage= @Vdd@}

){ Coupled{ nmos1.poisson nmos1.electron nmos1. nmos1.contact

nmos2.poisson nmos2.electron nmos2. nmos2.contact

nmos3.poisson nmos3.electron nmos3. nmos3.contact

nmos4.poisson nmos4.electron nmos4. nmos4.contact

pmos1.poisson pmos1.hole pmos1. pmos1.contact

pmos2.poisson pmos2.hole pmos2. pmos2.contact

circuit }

}

NewCurrentFile

Quasistationary(

InitialStep=1e-3 Increment=1.2

MinStep=1e-12 MaxStep=0.05

Goal{ Parameter=vin.dc Voltage= @Vdd@ })

{ Coupled{ nmos1.poisson nmos1.electron nmos1. nmos1.contact

nmos2.poisson nmos2.electron nmos2. nmos2.contact

nmos3.poisson nmos3.electron nmos3. nmos3.contact

nmos4.poisson nmos4.electron nmos4. nmos4.contact

pmos1.poisson pmos1.hole pmos1. pmos1.contact

pmos2.poisson pmos2.hole pmos2. pmos2.contact

circuit }

}}

*** Note: As for the SDEVICE Code:

```
System{
Vsource_psetvdd (dd 0) { dc = 0.0 }
Vsource_psetvwl (T 0) { dc = 0.0 }
Vsource_psetvb (L 0) { dc = 0.0 }
Vsource_psetvbl (R 0) { dc = 0.0 }
Vsource_pset vin (VinL 0) { dc = 0.0 }
NMOS1 nmos1( "S"=0   "D"=VinR "G"=VinL   )
NMOS2 nmos2( "S"=0   "D"=VinL "G"=VinR   )
PMOS1 pmos1( "S"=dd "D"=VinR "G"=VinL   )
PMOS2 pmos2( "S"=dd "D"=VinL "G"=VinR   )
NMOS3 nmos3( "S"=VinR   "D"=L "G"=T   )
NMOS4 nmos4( "S"=R   "D"=VinL "G"=T   )
Plot "n@node@_sys_des.plt" (time() v(VinL) v(dd) v(T) v(L) v(R) v(VinR)
)}
```

----------- SRAM circuit scheme -------------------

The syntax of System {…} is hereby explained:

Declare:vdd, with two terminals on dd node and ground and dc=0V;

Declare:vwl, with two terminals on T node and ground and dc=0V;

Declare: vb, with two terminals on L node and ground and dc=0V;

Declare:vbl, with two terminals on R node and ground and dc=0V;

Declare:V_{in}, with two terminals on VinL node and ground and dc=0V;

Declare:nmos1, with S node connected to ground, D node connected to VinR, and G node connected to VinL;

Declare:nmos2, with S node connected to ground, D node connected to VinL, and G node connected to VinR;

Fig. 4.13 Schematic of
SRAM input voltage of
programming codes

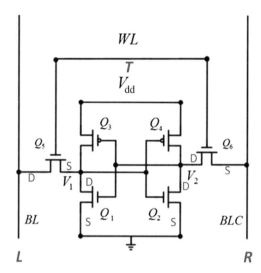

Declare:pmos1, with S node connected to dd, D node connected to VinR, and G
node connected to VinL;
Declare:pmos2, with S node connected to dd, D node connected to VinL, and G
node connected to VinR;
Declare:nmos3, with S node connected to VinR, D node connected to L, and G
node connected to T;
Declare:nmos4, with S node connected to R, D node connected to VinL, and G
node connected to T.

3. **INSPECT – dessis_des.cmd**

* ---------------------- **Ex 4.2 dessis_des.cmd** ----------------------*

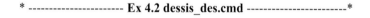

setout_filen@previous@_sys_des
proj_load "${out_file}.plt"
cv_createDS inv1 "${out_file} v(VinL)" "${out_file} v(VinR)"
cv_createDS inv2 "${out_file} v(VinR)" "${out_file} v(VinL)"

* --- END ---*

The figure of two cross-coupled Inverters in SRAM is as shown in Fig. 4.14,
which is known as butterfly curve. It reveals that the static noise margin (SNM) is
defined as the circumference of maximum rectangle between two voltage transfer
curves (with the unit of mV), which means that greater circumference of the square
will lead to stronger static noise margin (SNM) of SRAM. Therefore, the greater
value of SNM will contribute to the overall circuit performance. Practically,

SRAM of 3D FinFET

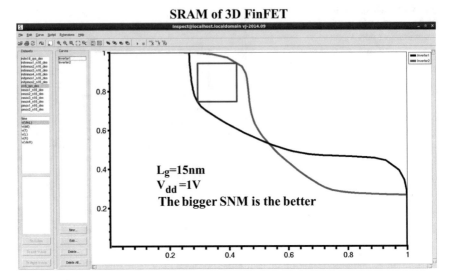

Fig. 4.14 Butterfly curve of SRAM based on L_g = 15 nm FinFET. The SNM is 120 mV

nFinFET and pFinFET must be perfectly matched in order to enhance the SNM as shown in Fig. 4.14, where $|V_{thp}| = |V_{thn}|$ and $I_{on,p} = I_{on,n}$, such that the static noise margin can also be enhanced. With continuous scaling of FinFET, the static noise margin has become more and more challenging.

In summary, the standard TCAD simulation examples of Inverter and SRAM with L_g = 15 nm FinFET as the fundamental transistor have been provided in this chapter. The numeric results of such simulation are in compliance with the current 14-nm/16-nm technology nodes of current semiconductor industry. Readers can understand the operating mechanisms of Inverter and SRAM based on these two examples, and they can serve as the reference for continuous scaling.

References

1. C.C. Hu, *Modern Semiconductor Devices for Integrated Circuits* (PERSON, 2010)
2. Y. Taur, T.H. Ning, *Fundamentals of Modern VLSI Device*, 2nd edn. (Cambridge University Press, New York, 2010)
3. C.H. Jan, F. Al-amoody, H.Y. Chang, T. Chang, Y.W. Chen, N. Dias, W. Hafez, D. Ingerly, M. Jang, E. Karl, S.K.Y. Shi, K. Komeyli, H. Kilambi, A. Kumar, K. Byon, C.G. Lee, J. Lee, T. Leo, P.C. Liu, N. Nidhi, R. Olac-vaw, C. Petersburg, K. Phoa, C. Prasad, C. Quincy, R. Ramaswamy, T. Rana, L. Rockford, A. Subramaniam, C. Tsai, P. Vandervoorn, L. Yang, A. Zainuddin, P. Bai, A 14 nm SoC platform technology featuring 2nd generation tri-gate transistors, 70 nm gate pitch, 52 nm metal pitch, and 0.0499 μm^2 SRAM cells, optimized for low power, high performance and high density SoC products. *VLSI Tech. Symp.* T12 (2015)

Chapter 5
Gate-All-Around (GAA) NWFET
with L_g = 10 nm Simulation

5.1 Introduction of Gate-All-Around Nanowire FET (GAA NWFET)

According to MOSFET scaling rule, the depletion layer formed in the channel of traditional 2D MOSFET near source and drain, the short-channel effect (SCE) has become inevitable along with the scaling of L_g dimension. As an important device design parameter, λ (nature length) relates to SCE, and it is depended to the geometric structure of device as shown in Fig. 5.1. Small λ value indicates that the channel is less vulnerable to the effect of depletion region in source and/or drain with biasing. Therefore, various multigate device structures have proposed in the industry for small λ value to reduce SCE, such as double-gate FET, tri-gate FET, FinFET. One of the superior solutions is gate-all-around nanowire FET (GAA NWFET or GAA FET) structure [1–6]. The evolution of MOSFET from planer, FinFET to GAA, is shown in Fig. 5.2. GAA FET could be the optimal solution based on previous description due to the smallest λ_0.

IBM introduced CMOS logic device and circuit performance of Si gate-all-around (GAA) nanowire MOSFET (NWFET) in 2013 [3], in which mentioned that the development of high performance CMOS logic technological application in the future would be heading toward GAA due to its excellent gate control capability, high I_{on}/I_{off} ratio, and extremely high density. Figure 5.3 shows 3D stacking GAA NWFET structure. It has three advantages: (1). Multiple NWs stacked based on the 3D stacking technology can lead to higher ON current, (2). Superior gate control capability can reduce SCE. (3). It is compatible to current CMOS FinFET technology. In short, GAA FET has superior performance and good candidate for next-generation technology. In Fig. 5.4, 2015 ITRS version 2.0 predicts GAA will apply after the year 2024.

Following section, we will discuss the design guideline and simulation of 3D GAA NWFET.

© Springer Nature Singapore Pte Ltd. 2018
Y.-C. Wu and Y.-R. Jhan, *3D TCAD Simulation for CMOS Nanoeletronic Devices*,
DOI 10.1007/978-981-10-3066-6_5

Single gate	$\lambda_1 = \sqrt{\dfrac{\varepsilon_{si}}{\varepsilon_{ox}} t_{si} t_{ox}}$
Double gate	$\lambda_2 = \sqrt{\dfrac{\varepsilon_{si}}{2\varepsilon_{ox}} t_{si} t_{ox}}$
Trigate (FinFET)	$\lambda_3 = \sqrt{\dfrac{\varepsilon_{si}}{3\varepsilon_{ox}} t_{si} t_{ox}}$
Quadruple gate	$\lambda_4 \cong \sqrt{\dfrac{\varepsilon_{si}}{4\varepsilon_{ox}} t_{si} t_{ox}}$
Surrounding gate (GAA)	$\lambda_0 = \sqrt{\dfrac{2\varepsilon_{si} t^2_{si} \ln(1+\dfrac{2t_{ox}}{t_{si}}) + \varepsilon_{ox} t^2_{si}}{16\varepsilon_{ox}}}$

Fig. 5.1 Natural length in devices with different gate structure

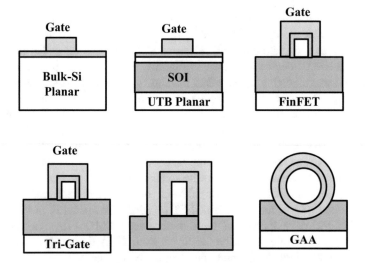

Fig. 5.2 Device structure evolution of MOSFET

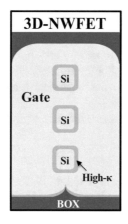

Fig. 5.3 Schmatic plot of 3D-NWFET with stacked nanowires

YEAR OF PRODUCTION	2015	2017	2019	2021	2024	2027	2030
Logic device technology naming	P70M56	P48M36	P42M24	P32M20	P24M12G1	P24M12G2	P24M12G3
Logic industry "Node Range" Labeling (nm)	"16/14"	"11/10"	"8/7"	"6/5"	"4/3"	"3/2.5"	"2/1.5"
Logic device structure options	FinFET FDSOI	FinFET FDSOI	FinFET LGAA	FinFET LGAA VGAA	VGAA,M3D	VGAA,M3D	VGAA,M3D
LOGIC DEVICE GROUND RULES							
MPU/SoC Metalx $^1/_2$ Pitch (nm)	28.0	18.0	12.0	10.0	6.0	6.0	6.0
MPU/SoC Metal0/1 $^1/_2$ Pitch (nm)	28.0	18.0	12.0	10.0	6.0	6.0	6.0
L_g : Physical Gate Length for HP Logic (nm)	24	18	14	10	10	10	10
L_g : Physical Gate Length for LP Logic (nm)	26	20	16	12	12	12	12
FinFET Fin Width (nm)	8.0	6.0	6.0	NA	N/A	N/A	N/A
FinFET Fin Height (nm)	42.0	42.0	42.0	NA	N/A	N/A	N/A
Device effective width - [nm]	92.0	90.0	56.5	56.5	56.5	56.5	56.5
Device lateral half pitch (nm)	21.0	18.0	12.0	10.0	6.0	6.0	6.0
Device width or diameter (nm)	8.0	6.0	6.0	6.0	5.0	5.0	5.0
DEVICE PHYSICAL&ELECTRICAL SPECS							
Power Supply Voltage - V_{dd} (V)	0.80	0.75	0.70	0.65	0.55	0.45	0.40
Subthreshold slope - [mV/dec]	75	70	68	65	40	25	25
Inversion layer thickness - [nm]	1.10	1.00	0.90	0.85	0.80	0.80	0.80
$V_{t,sat}$ (mV) at I_{off}=100nA/um - HP Logic	129	129	133	136	84	52	52
$V_{t,sat}$ (mV) at I_{off}=100pA/um - LP Logic	351	336	333	326	201	125	125
Effective mobility (cm2/V.s)	200	150	120	100	100	100	100
Rext (Ohms.um) - HP Logic [7]	280	238	202	172	146	124	106
Ballisticity.Injection velocity (cm/s)	1.20E-07	1.32E-07	1.45E-07	1.60E-07	1.76E-07	1.93E-07	2.13E-07
V_{dsat} (V) - HP Logic	0.115	0.127	0.136	0.128	0.141	0.155	0.170
V_{dsat} (V) - LP Logic	0.125	0.141	0.155	0.153	0.169	0.186	0.204
I_{on} (uA/um) at I_{off}=100nA/um - HP logic w/ Rext=0	2311	2541	2782	2917	3001	2670	2408

Fig. 5.4 Selected logic core device technology road map as predicted by 2015 ITRS version 2.0 [1, 2]

5.2   [Example 5.1] 3D IM n-Type GAA NWFET

The following three main program code files are based on Synopsys Sentaurus TCAD 2014 version. For the simulation tools detailed content of **SDE → devise_dvs.cmd, SDEVICE → dessis_des.cmd, and INSPECT → inspect_inc. cmd**, are identical to Example 3.1.

In this section, we only introduce the SDE tool codes (Figs. 5.5 and 5.6).

3D – n-type GAA NWFET Mesh

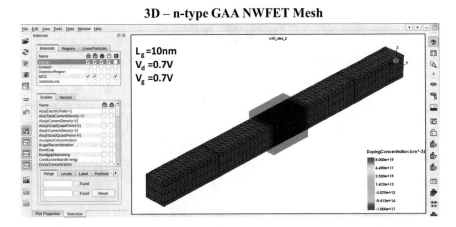

Fig. 5.5 Mesh diagram of 3D n-type GAA NWFET simulation

Fig. 5.6 The required simulation tools shown in the workbench of n-type GAA NWFET simulation

1. **SDE → devise_dvs.cmd**

```
;------------------- Exapmle 5.1 nGAAFET Lg=10nm ----------------;
;---------------------------------- parameter -------------------------------;
(define W 5)
(define tox 1)
(define Lg @Lg@)
(define LSDC 15)
(define LSD 15)
(define C_Doping 1e17)
(define SD_Doping 8e19)
(define nm 1e-3)
(define x1 LSDC)
(define x2 (+ x1 LSD))
(define x3 (+ x2 Lg))
(define x4 (+ x3 LSD))
(define x5 (+ x4 LSDC))
(define y1 tox)
(define y2 (+ y1 W))
(define y3 (+ y2 tox))
(define z1 tox)
(define z2 (+ z1 W))
(define z3 (+ z2 tox))
;---------------------------------- Structure -----------------------------------;
;--- Source ---;
"ABA"
(sdegeo:create-cuboid (position 0 y1 z1 )  (position x1 y2 z2 )  "Silicon"
"SourceC")
(sdegeo:create-cuboid (position x1 y1 z1 )  (position x2 y2 z2 )  "Silicon" "Source")
;--- Gate oxide ---;
(sdegeo:create-cuboid (position x2 0 0 )  (position x3 y3 z3 )  "SiO2" "Gateoxide")
;--- Channel ---;
  (sdegeo:create-cuboid (position x2 y1 z1 )  (position x3 y2 z2 )  "Silicon"
"Channel")
```

```
;--- Drain ---;
(sdegeo:create-cuboid (position x3 y1 z1 )   (position x4 y2 z2 )   "Silicon" "Drain")
(sdegeo:create-cuboid (position x4 y1 z1 )   (position x5 y2 z2 )   "Silicon"
"DrainC")
;------------------------------------- Contact -------------------------------------;
;----- Gate -----;
(sdegeo:define-contact-set "G" 4.0   (color:rgb 1.0 0.0 0.0 ) "||" )
(sdegeo:set-current-contact-set "G")
(sdegeo:set-contact-faces
(find-face-id (position (+ x2 1) (+ y1 1) z3 )))
(sdegeo:define-contact-set "G" 4.0   (color:rgb 1.0 0.0 0.0 ) "||" )
(sdegeo:set-current-contact-set "G")
(sdegeo:set-contact-faces
(find-face-id (position (+ x2 1) (+ y1 1) 0 )))
(sdegeo:define-contact-set "G" 4.0   (color:rgb 1.0 0.0 0.0 ) "||" )
(sdegeo:set-current-contact-set "G")
(sdegeo:set-contact-faces
(find-face-id (position (+ x2 1) 0 1 )))
(sdegeo:define-contact-set "G" 4.0   (color:rgb 1.0 0.0 0.0 ) "||" )
(sdegeo:set-current-contact-set "G")
(sdegeo:set-contact-faces
(find-face-id (position (+ x2 1)   y3 1 )))
;----- Drain -----;
(sdegeo:define-contact-set "D" 4.0   (color:rgb 1.0 0.0 0.0 ) "##" )
(sdegeo:set-current-contact-set "D")
(sdegeo:set-contact-faces
  (find-face-id (position (+ x4 (/ LSDC 2)) (- y2 1) z2 )))
;----- Source -----;
(sdegeo:define-contact-set "S" 4.0   (color:rgb 1.0 0.0 0.0 ) "##" )
(sdegeo:set-current-contact-set "S")
(sdegeo:set-contact-faces
(find-face-id (position (- x1 1) (- y2 1) z2 )))
```

```
;------------------------------------- Doping -------------------------------------;
;----- Channel -----;
(sdedr:define-constant-profile "dopedC" "BoronActiveConcentration" C_Doping )
(sdedr:define-constant-profile-region    "RegionC" "dopedC" "Channel" )
;----- Source -----;
(sdedr:define-constant-profile "dopedS" "ArsenicActiveConcentration" SD_Doping )
(sdedr:define-constant-profile-region    "RegionS" "dopedS" "Source" )
(sdedr:define-constant-profile "dopedSC" "ArsenicActiveConcentration"
SD_Doping )
(sdedr:define-constant-profile-region    "RegionSC" "dopedSC" "SourceC" )
;----- Drain ------;
(sdedr:define-constant-profile "dopedD" "ArsenicActiveConcentration" SD_Doping )
(sdedr:define-constant-profile-region    "RegionD" "dopedD" "Drain" )
(sdedr:define-constant-profile "dopedDC" "ArsenicActiveConcentration"
SD_Doping )
(sdedr:define-constant-profile-region    "RegionDC" "dopedDC" "DrainC" )



;------------------------------------- Mesh -------------------------------------;
;--- AllMesh ---;
(sdedr:define-refinement-size "Cha_Mesh" 2 2 2 2 2 2)
(sdedr:define-refinement-material "channel_RF" "Cha_Mesh" "Silicon" )
;--- ChannelMesh ---;
(sdedr:define-refinement-window "multiboxChannel" "Cuboid"
(position 30 y1 z1)
(position (+ 30 Lg) y2 z2)    )
(sdedr:define-multibox-size "multiboxSizeChannel"    2 2 2 0.5 0.5 0.5)
(sdedr:define-multibox-placement "multiboxPlacementChannel"
"multiboxSizeChannel" "multiboxChannel")
(sdedr:define-refinement-function "multiboxPlacementChannel"
"DopingConcentration" "MaxTransDiff" 1)
```

```
;----------------    Save BND and CMD and rescale to nm   --------------------;
(sde:assign-material-and-region-names (get-body-list) )
(sdeio:save-tdr-bnd (get-body-list) "n@node@_nm.tdr")
(sdedr:write-scaled-cmd-file "n@node@_msh.cmd" nm)
(define sde:scale-tdr-bnd
(lambda (tdrin sf tdrout)
(sde:clear)
(sdegeo:set-default-boolean "XX")
(sdeio:read-tdr-bnd tdrin)
(entity:scale (get-body-list) sf)
(sdeio:save-tdr-bnd (get-body-list) tdrout)
   ))
(sde:scale-tdr-bnd "n@node@_nm.tdr" nm "n@node@_bnd.tdr")
#------------------------------------- END ----------------------------------------#
```

The electrical property I_d–V_g of simulation result is shown in Figs. 5.7 and 5.8, with important parameters of

1. SS at around 72 mV/dec.,
2. V_{tn} at around 0.32 V,
3. I_{sat} at around 7.2×10^{-6} A, and

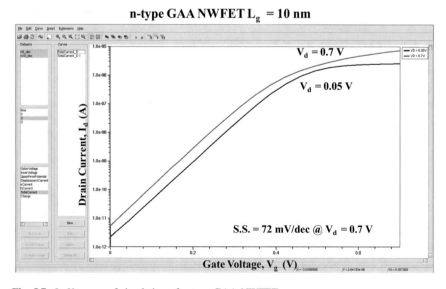

Fig. 5.7 I_d–V_g curve of simulation of n-type GAA NWFET

Vd		VT	gmax	ioff	isat	sslop	rout
0.05	--	0.3556467984369356	1.413999375389927e-05	2.26559632532391e-12	2.54114151449987e-06	0.07552239720673501	14984983.98499136
0.7	--	0.3206659157134897	2.819217653455339e-05	5.52808186759873e-12	7.20077248708731e-06	0.07242815780285297	5690002.43255516

Fig. 5.8 Electrical property parameters of n-type GAA NWFET simulation

3D – Electron concentration

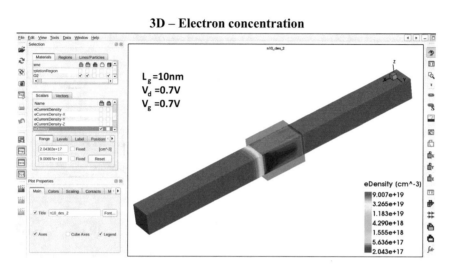

Fig. 5.9 Electron concentration distribution of 3D n-type GAA NWFET simulation

2D – Electron concentration

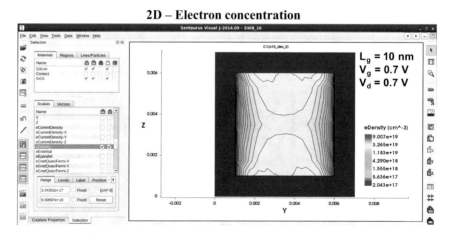

Fig. 5.10 Electron concentration distribution of 2D n-type GAA NWFET simulation

3D-Electric-Field

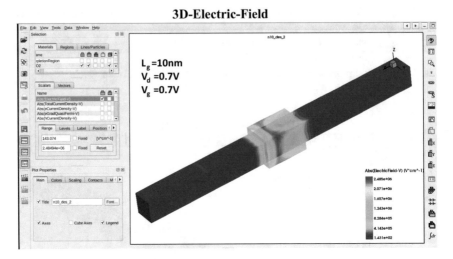

Fig. 5.11 Electrical field distribution of 3D n-type GAA NWFET simulation

4. I_{off} at around 5.52×10^{-12} A.

As shown in Fig. 5.6, the simulation conditions are $L_g = 10$ nm, $V_d = 0.7$, 0.05 V, and $V_g = 0.7$ V.

The next is the discussion of physical property analysis of 3D n-type GAA NWFET. The structural channel mesh, the electron concentration distributions of 3D and 2D structures, electrical field distributions, electrical potential distributions, and the energy band diagrams along the channel direction are shown in Figs. 5.9, 5.10, 5.11, 5.12, 5.13, 5.14, 5.15, 5.16, and 5.17 with the conditions of

2D-Electric-Field

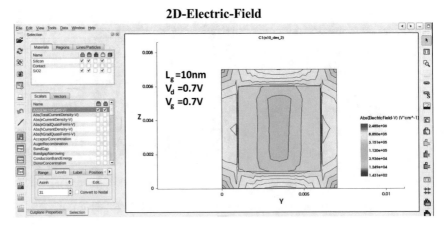

Fig. 5.12 Electrical field distribution of 2D n-type GAA NWFET simulation

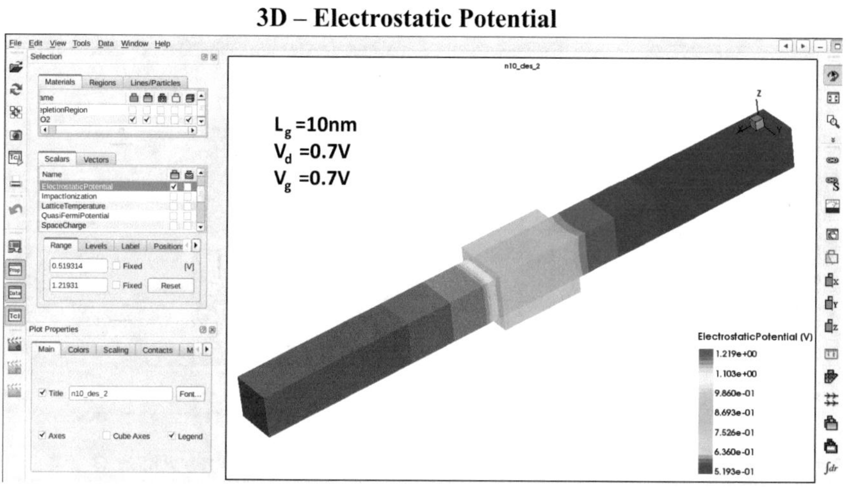

3D – Electrostatic Potential

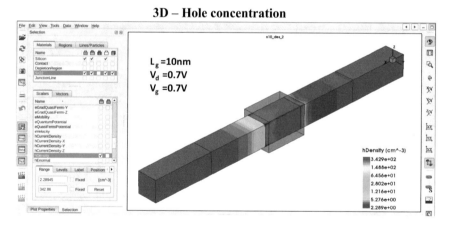

Fig. 5.13 Electrical potential distribution of 3D n-type GAA NWFET simulation

3D – Hole concentration

Fig. 5.14 Hole concentration distribution of 3D n-type GAA NWFET simulation

$L_g = 10$ nm, $V_d = 0.7$, 0.05 V and $V_g = 0.7$ V, respectively. The reader can follow the following steps to analyze the physical property of 3D n-type GAA NWFET.

The electron concentration distribution of 3D n-type GAA NWFET simulation in Fig. 5.9 can be converted into 2D by cutting along X-axis, which is the Y-Z cross section, as shown in Fig. 5.10 as the channel cross section.

2D – Hole concentration

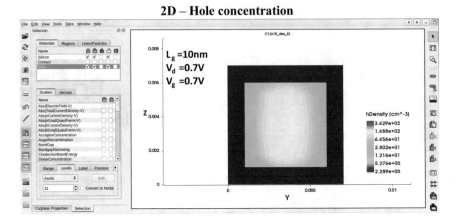

Fig. 5.15 Hole concentration distribution of 2D n-type GAA NWFET simulation

3D – Electron Current Density

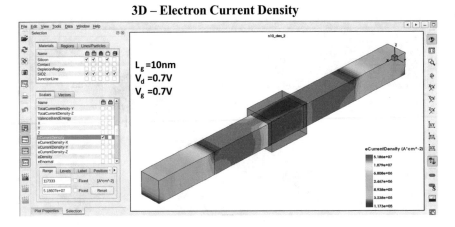

Fig. 5.16 Electron current density distribution of 3D n-type GAA NWFET simulation

5.3 [Example 5.2] 3D IM p-Type GAA NWFET

The following three main program code files are based on Synopsys Sentaurus TCAD 2014 version. For the simulation tools detailed content of **SDE → devise_dvs.cmd,** **SDEVICE → dessis_des.cmd,** and **INSPECT → inspect_inc.cmd**. Please refer to Examples 3.1 and 5.1.

In this section, we can introduce the **SDE tool codes in doping section.**

Band Diagram

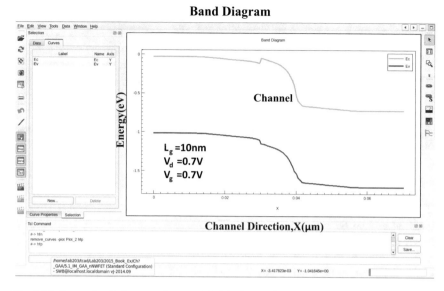

Fig. 5.17 Energy band diagram of n-type GAA NWFET simulation

;------------   Exapmle 5.2 pGAAFET Lg=10nm   ---------;

;----- Channel -----;

(isedr:define-constant-profile "dopedC" "ArsenicActiveConcentration"
C_Doping)

(isedr:define-constant-profile-region   "RegionC" "dopedC" "Channel")

;----- Source -----;

(isedr:define-constant-profile "dopedS" "BoronActiveConcentration" SD_Doping)

(isedr:define-constant-profile-region   "RegionS" "dopedS" "Source")

(isedr:define-constant-profile "dopedSC" "BoronActiveConcentration"
SD_Doping)

(isedr:define-constant-profile-region   "RegionSC" "dopedSC" "SourceC")

;----- Drain ------;

(isedr:define-constant-profile "dopedD" "BoronActiveConcentration" SD_Doping)

(isedr:define-constant-profile-region   "RegionD" "dopedD" "Drain")

(isedr:define-constant-profile "dopedDC" "BoronActiveConcentration"
SD_Doping)

(isedr:define-constant-profile-region   "RegionDC" "dopedDC" "DrainC")

;------------------------------------- END-------------------------------------;

Fig. 5.18 The required simulation tools are shown in the workbench of p-type GAA NWFET simulation

The I_d–V_g curve of simulation result is shown in Fig. 5.18 with important parameters of

1. SS at around 69 mV/dec.,
2. V_{tp} at around −0.32 V,
3. I_{sat} at around 5.35 × 10^{-6} A, and
4. I_{off} at around 4.93 × 10^{-12} A.

The results are from Fig. 5.19 with conditions of L_g = 10 nm, V_d = −0.7, −0.05 V, and V_g = −0.7 V (Fig. 5.20).

The next is the discussion of physical property analysis of 3D p-type GAA NWFET. The structural channel mesh, the electron concentration distributions of 3D and 2D structures, electrical field distributions, electrical potential distributions, and the energy band diagrams along the channel direction are shown

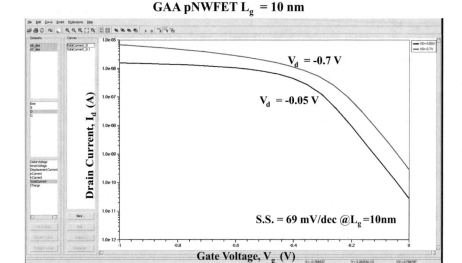

Fig. 5.19 I_d–V_g curve of p-type GAA NWFET simulation

Vd		gmax	ioff	isat	sslop	rout	VT
-0.7	--	-1.626619838658618e-10	4.93237804172092e-12	5.35051415326732e-06	0.06970613421300044	-48.31188817188574	-0.3254946847406228
-0.05	--	-1.123263384424954e-10	4.20731719311848e-12	1.7251840667619e-06	0.080079021928176	-122.1562673174145	-0.359572246765028

Fig. 5.20 Electrical property parameters of p-type GAA NWFET simulation

3D –p-type GAA NWFET Mesh

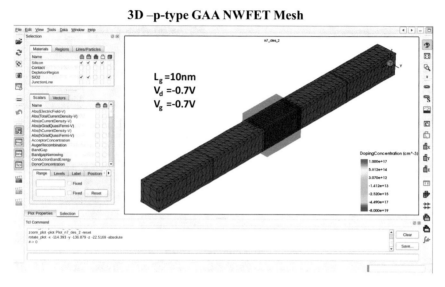

Fig. 5.21 Mesh diagram of p-type GAA NWFET simulation

3D – p-type GAA NWFET Mesh (with SiO$_2$)

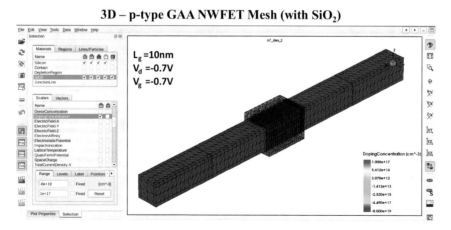

Fig. 5.22 Mesh diagram (including SiO$_2$) of 3D p-type GAA NWFET simulation

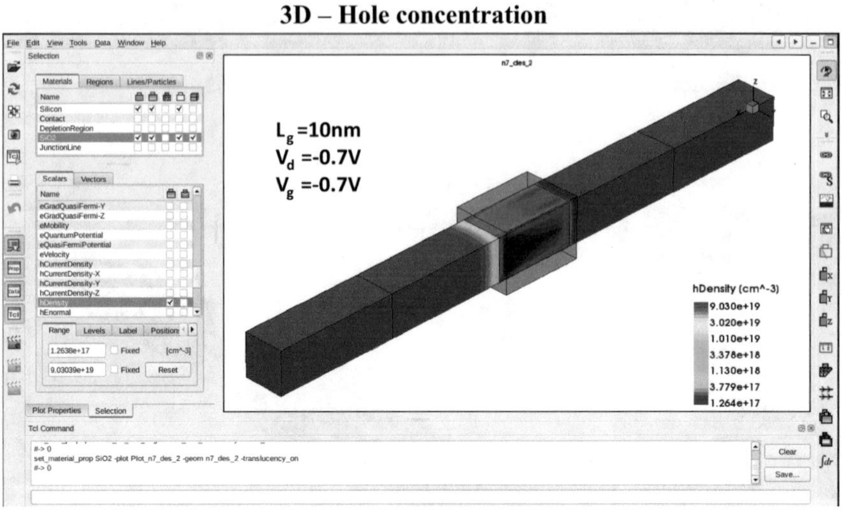

3D – Hole concentration

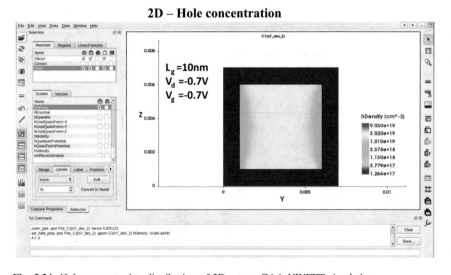

Fig. 5.23 Hole concentration distribution of 3D p-type GAA NWFET simulation

2D – Hole concentration

Fig. 5.24 Hole concentration distribution of 2D p-type GAA NWFET simulation

in Figs. 5.21, 5.22, 5.23, 5.24, 5.25, 5.26, 5.27, 5.28, 5.29, 5.30, and 5.31 with the
conditions of L_g = 10 nm, V_d = −0.7, −0.05 V, and V_g = −0.7 V, respectively.
The reader can follow the following steps to analyze the physical property of 3D
p-type GAA NWFET.

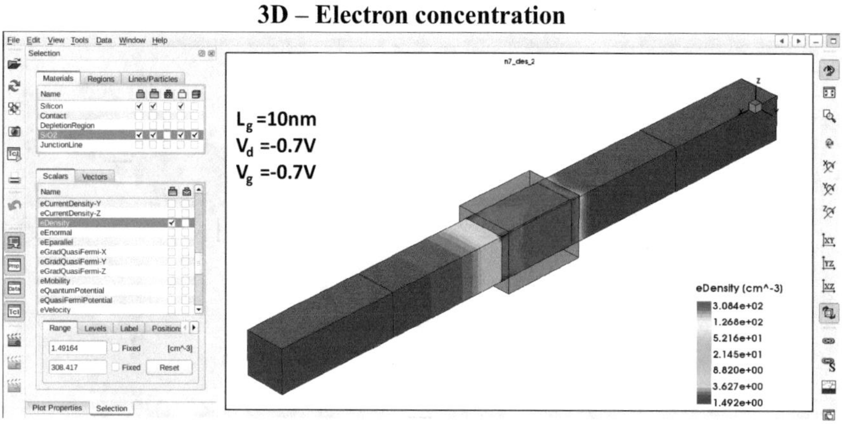

3D – Electron concentration

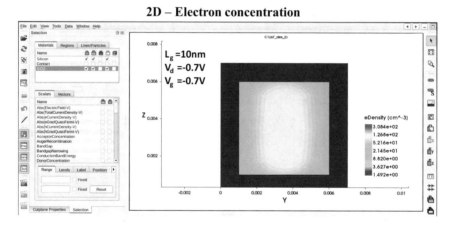

Fig. 5.25 Electron concentration distribution of 3D p-type GAA NWFET simulation

2D – Electron concentration

Fig. 5.26 Electron concentration distribution of 2D p-type GAA NWFET simulation

5.4 [Example 5.3] 3D Cylindrical IM n-Type GAA NWFET

The best symmetrical gate-all-around (GAA) structure is cylindrical NWFET. This example simulates the inversion-mode (IM) GAA FET with $L_g = 10$ nm, radius ($r = 3, 4, 5$ nm), and gate insulator using $HfO_2 = 2$ nm.

Simulation tools detailed content of **SDE → devise_dvs.cmd, SDEVICE → dessis_des.cmd, and INSPECT → inspect_inc.cmd**, are identical to Example 3.1 nFinFET. In this section, we only introduce the **SDE tool codes in doping section** (Figs. 5.32, 5.33, 5.34, 5.35, and 5.36).

3D – Hole Current Density

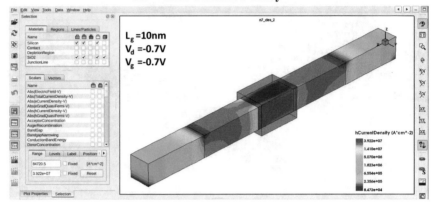

Fig. 5.27 Hole current density distribution of 3D p-type GAA NWFET simulation

2D – Hole Current Density

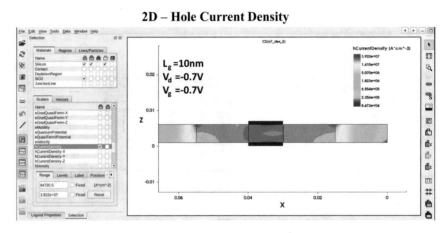

Fig. 5.28 Hole current density distribution of 2D p-type GAA NWFET simulation

3D-Electric-Field

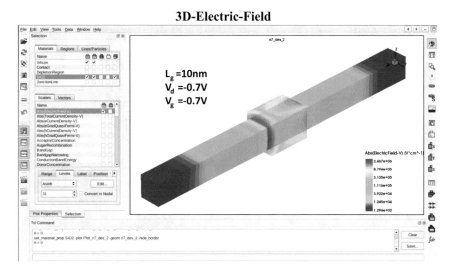

Fig. 5.29 Electrical field distribution of 3D p-type GAA NWFET simulation

2D-Electric-Field

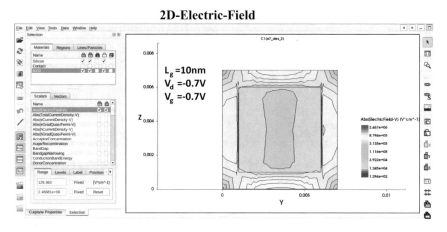

Fig. 5.30 Electrical field distribution of 2D p-type GAA NWFET simulation

Band Diagram

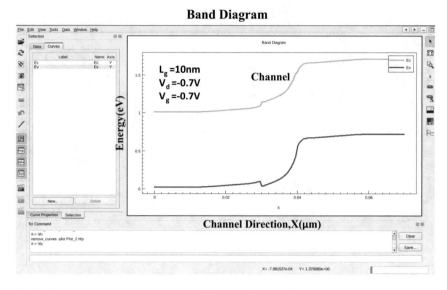

Fig. 5.31 Energy band diagram of p-type GAA NWFET simulation

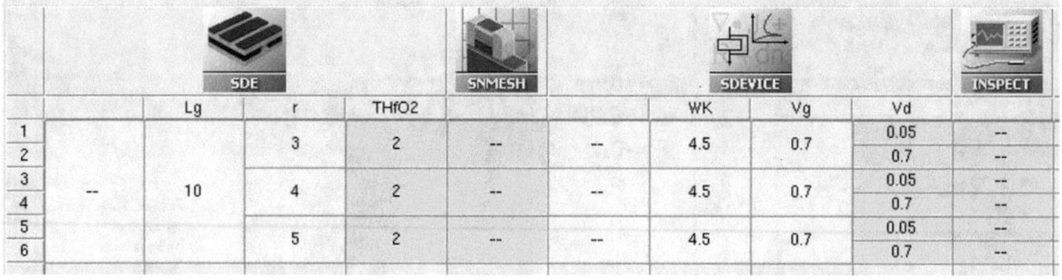

	Lg	r	THfO2			WK	Vg	Vd	
1		3	2	--	--	4.5	0.7	0.05	--
2								0.7	--
3	--	10	4	2	--	--	4.5	0.7	0.05	--
4								0.7	--
5		5	2	--	--	4.5	0.7	0.05	--
6								0.7	--

Fig. 5.32 Simulation device parameters of 3D cylindrical IM n-type GAA NWFET with different radius $r = 3$, 4, and 5 nm and $HfO_2 = 2$ nm

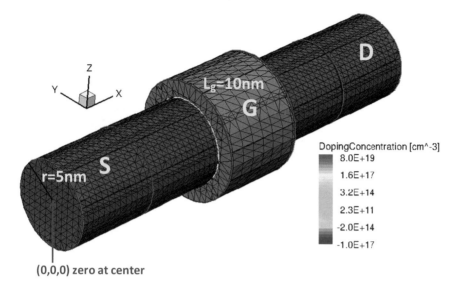

Fig. 5.33 Simulation device structure of 3D cylindrical IM n-type GAA NWFET

	VT	gmax	ioff	isat	sslop	rout
--	0.306156	2.9897674e-05	3.7391454e-12	4.6464688e-06	0.069112716	8378323.7
--	0.28954774	5.6789086e-05	2.4633824e-12	1.6541364e-05	0.058969546	10936031
--	0.27577754	4.2257918e-05	3.1911838e-11	7.8129734e-06	0.079381445	1099703.5
--	0.24015506	8.5391831e-05	3.2215939e-11	2.5416133e-05	0.065602547	928902.32
--	0.23698386	5.8330731e-05	2.4711434e-10	1.1714029e-05	0.091941023	162214.07
--	0.17556371	0.00011806855	6.4792287e-10	3.6227389e-05	0.076222804	51967.558

Fig. 5.34 Electrical property parameters of p-type GAA NWFET simulation

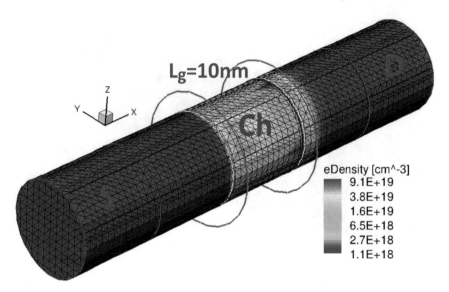

Fig. 5.35 Electron density of parameters of 3D cylindrical IM n-type GAA NWFET

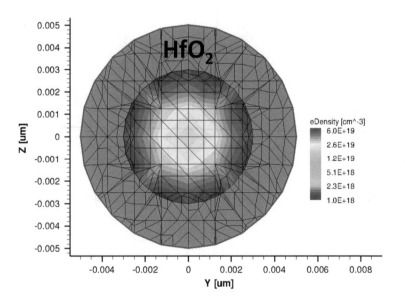

Fig. 5.36 Electron current density of 2D Y–X channel cross section of 3D cylindrical IM n-type GAA NWFET with radius (r) of 3 nm

```
;-----Example 5.3 3D cylindrical IM n-type GAA NWFET Lg=10nm   ----------------;
;---------------------------------- parameter ----------------------------------------;
(define r @r@)
(define THfO2 @THfO2@)
(define Lg @Lg@)
(define LSDC 10)
(define LSD 10)
(define C_Doping 1e17)
(define SD_Doping 8e19)
(define nm 1e-3)
(define x1 LSDC)
(define x2 (+ x1 LSD))
(define x3 (+ x2 Lg))
(define x4 (+ x3 LSD))
(define x5 (+ x4 LSDC))


;----------------------------------- Structure ------------------------------------------;
(sdegeo:create-cylinder (position 0 0 0) (position x1 0 0)   r "Silicon" "SourceC")
(sdegeo:create-cylinder (position x1 0 0) (position x2 0 0) r "Silicon" "Source")
(sdegeo:create-cylinder (position x2 0 0) (position x3 0 0) r "Silicon" "Channel" )
(sdegeo:set-default-boolean "BAB")
(sdegeo:create-cylinder (position x2 0 0) (position x3 0 0) (+ r THfO2) "HfO2"
"GateOxide" )
(sdegeo:create-cylinder (position x3 0 0) (position x4 0 0) r "Silicon" "Drain")
(sdegeo:create-cylinder (position x4 0 0) (position x5 0 0) r "Silicon" "DrainC")


;--------------------------------------- Doping ----------------------------------------;
;----- Channel -----;
(sdedr:define-constant-profile "dopedC" "BoronActiveConcentration" C_Doping )
(sdedr:define-constant-profile-region   "RegionC" "dopedC" "Channel" )
;----- Source -----;
(sdedr:define-constant-profile "dopedS" "ArsenicActiveConcentration" SD_Doping )
(sdedr:define-constant-profile-region   "RegionS" "dopedS" "Source" )
(sdedr:define-constant-profile "dopedSC" "ArsenicActiveConcentration"
SD_Doping )
(sdedr:define-constant-profile-region   "RegionSC" "dopedSC" "SourceC" )
```

```
;----- Drain ------;
(sdedr:define-constant-profile "dopedD" "ArsenicActiveConcentration" SD_Doping )
(sdedr:define-constant-profile-region   "RegionD" "dopedD" "Drain" )
(sdedr:define-constant-profile "dopedDC" "ArsenicActiveConcentration"
SD_Doping )
(sdedr:define-constant-profile-region   "RegionDC" "dopedDC" "DrainC" )
;------------------------------------- Contact -------------------------------------;
;----- Source -----;
(sdegeo:define-contact-set "S" 4.0    (color:rgb 1.0 0.0 0.0 ) "##" )
(sdegeo:set-current-contact-set "S")
(sdegeo:define-3d-contact(list(car(find-face-id (position (/ x1 2) r 0    ))))"S")
;----- Gate -----;
(sdegeo:define-contact-set "G" 4.0    (color:rgb 1.0 0.0 0.0 ) "##" )
(sdegeo:set-current-contact-set "G")
(sdegeo:define-3d-contact(list(car(find-face-id (position (+ x2 (/ Lg 2)) (+ r THfO2)
0    ))))"G")
;----- Drain -----;
(sdegeo:define-contact-set "D" 4.0    (color:rgb 1.0 0.0 0.0 ) "##" )
(sdegeo:set-current-contact-set "D")
(sdegeo:define-3d-contact(list(car(find-face-id (position (+ x4 (/ LSDC 2)) r
0    ))))"D")


;------------------------------------- Mesh -------------------------------------;
;--- AllMesh ---;
(sdedr:define-refinement-size "Cha_Mesh" 2 2 2 1 1 1)
(sdedr:define-refinement-material "channel_RF" "Cha_Mesh" "Silicon" )
```

```
;--- ChannelMesh ---;
;(sdedr:define-refinement-window "multiboxChannel" "Cuboid"
;   (position x1 r r)
;   (position x4 (- r) (- r))   )
;(sdedr:define-multibox-size "multiboxSizeChannel" 1 1 1 1 1 1)
;(sdedr:define-multibox-placement "multiboxPlacementChannel"
"multiboxSizeChannel" "multiboxChannel")
(sdedr:define-refinement-function "multiboxPlacementChannel"
"DopingConcentration" "MaxTransDiff" 1)
;---------------------- Save BND and CMD and rescale to nm ---------------------;
(sde:assign-material-and-region-names (get-body-list) )
(sdeio:save-tdr-bnd (get-body-list) "n@node@_nm.tdr")
(sdedr:write-scaled-cmd-file "n@node@_msh.cmd" nm)
(define sde:scale-tdr-bnd
(lambda (tdrin sf tdrout)
(sde:clear)
(sdegeo:set-default-boolean "XX")
(sdeio:read-tdr-bnd tdrin)
(entity:scale (get-body-list) sf)
(sdeio:save-tdr-bnd (get-body-list) tdrout)
))
(sde:scale-tdr-bnd "n@node@_nm.tdr" nm "n@node@_bnd.tdr")
;------- END----:
```

The simulation results are shown in following figure.

According to Fig. 5.37, the best performance with highest I_{on}/I_{off} and smallest SS of cylindrical GAA NWFET is radius of 3 nm. For higher I_{on}, we can use vertical stacked multilayer cylindrical GAA NWFET (Fig. 5.3).

Summary of this chapter: The best 3D TCAD simulation example of GAA NWFET has been provided for nFET and pFET with L_g = 10 nm. GAA is equipped with better gate control capability than FinFET, while it also can easily apply in 3D-IC technology such as the vertical GAA structure. It has the potential for continuous scaling down to sub-5-nm technology node, which can be an excellent candidate for future high performance, CMOS logic technology applications.

Fig. 5.37 Electrical I_d–V_g
property parameters of p-type
GAA NWFET simulation
with different radius (r)

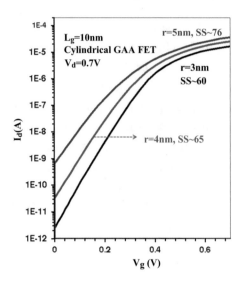

References

1. J.P. Colinge, Multiple-gate SOI MOSFETs. Solid-State Electron. **48**, 875 (2004)
2. N. Singh, K.D. Buddharaju, S.K. Manhas, A. Agarwal, S.C. Rustagi, G.Q. Lo, N. Balasubramanian, D.L. Kwong, Si, SiGe nanowire devices by top–down technology and their applications. IEEE Trans. Electron Devices **55**, 3107 (2008)
3. K. Nayak, M. Bajaj, A. Konar, P.J. Oldiges, K. Natori, H. Iwai, K.V.R.M. Murali, V.R. Rao, CMOS logic device and circuit performance of si gate all around nanowire MOSFET. IEEE Trans. Electron Devices **61**, 3066 (2014)
4. ITRS version 2.0 (2015), http://www.semiconductors.org/main/2015_international_technology_roadmap_for_semiconductors_itrs/
5. M.S. Yeh, Y.J. Lee, M.F. Hung, K.C. Liu, Y.C. Wu, High-performance Gate-all-around poly-Si thin-film transistors by microwave annealing with NH_3 plasma passivation. IEEE Trans. Nanotechnol. **12**, 636 (2013)
6. H.B. Chen, C.Y. Chang, N.H. Lu, J.J. Wu, M.H. Han, Y.C. Cheng, Y.C. Wu, Characteristics of Gate-all-around junctionless poly-Si TFTs with an ultrathin channel. IEEE Electron Device Lett. **34**, 897 (2013)

Chapter 6
Junctionless FET with L_g = 10 nm Simulation

6.1 Foreword

There will be junctions in the channels of source and drain of traditional inversion-mode (IM) MOSFET or FinFET, such as the nFET with doping style of N^+PN^+. It is difficult for the ion implantation process to generate ultra-shallow junction with high concentration due to diffusion and semiconductor dopant characteristic. In addition, the substantial difference among the dopant concentrations of channel, source, and drain will result in the PN depletion region, thus causing the effective channel length of semiconductor device to be shorter than the length defined by the lithography. It is a significant challenge for the scaling of semiconductor device feature size to overcome the two aforementioned phenomena. The doping type and concentration are identical of source, channel, and drain. The n-type JL—FET will not suffer from the issues of PN junction and dopant concentration gradient, and it is based on a relatively simple process with low thermal budget, making it feasible for gate first process. Therefore, it is filled with potential for the application of future sub-10 nm technology node [1–7].

Without any extra bias, the traditional inversion-mode MOSFET will rely on high PN junction energy barrier in the channel to stop the current conduction, thus achieving rather low off current ($I_{off} < 1$ nA/μm). However, the channel, source, and drain of JL—FET are based on same polarity and doping of same concentration, such as n-type JL—FET with doping of $N^+N^+N^+$. Therefore, the channel cannot generate any energy barrier to inhibit the current conduction. That is why JL—FET is a normally on device. There are two methods for making JL—FET normally off: (1) raising energy barrier of channel based on the difference between work functions (WF) of metal gate and Si channel, and (2) making Si channel fully depleted by ultra-thin channel's thickness [6, 7].

© Springer Nature Singapore Pte Ltd. 2018
Y.-C. Wu and Y.-R. Jhan, *3D TCAD Simulation for CMOS Nanoeletronic Devices*,
DOI 10.1007/978-981-10-3066-6_6

6.2 Short-Channel Effect (SCE) of CMOS Device

The diagrams of JL—FET and IM-FET are as shown in Fig. 6.1a, b. As for the traditional IM-FET, with voltage applied to drain, the junction of drain and channel is at reverse bias and the depletion region is widening, thus causing the effective gate length (L_g) to be reduced. This is the so-called SCE. In addition, drain voltage will also reduce the energy barrier of channel, which is known as drain-induced barrier lowing (DIBL). These two aforementioned phenomena are not obvious in the MOSFET with long channel. However, along with the scaling of channel length, these two phenomena will cause V_{th} to be reduced, which is known as V_{th} roll-off based on the equation as shown below:

$$V_{th} = V_{tho} - SCE - DIBL \tag{6.1}$$

where V_{tho} is the V_{th} of the channel with long L_g.

It is shown in Fig. 6.2a that effective channel length is greater than physical length ($L_{eff} > L_{physical}$) when n-channel JL—FET is in off state, and this phenomenon can greatly improve short channel effect (SCE). The situation of n-channel JL—FET in on-state is as shown in Fig. 6.2b. In Fig. 6.2c, n-channel IM—FET with the effective gate length is L_{eff} when the device is on, and the effective gate length is L_{SCE} when the device is off. L_{SCE} is defined as the length of the (p-type) neutral region in the channel. It is smaller than $L_{physical}$ because of the depletion regions from the source and drain PN junctions in the IM—FET channel region [6].

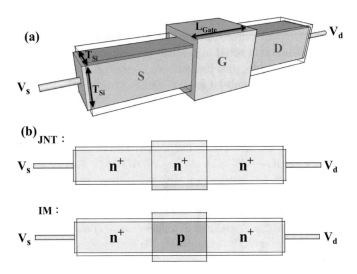

Fig. 6.1 a Aerial views of JL—FET (JNT) and IM-FET (IM), **b** longitudinal concentration distribution of JNT and IM [5]

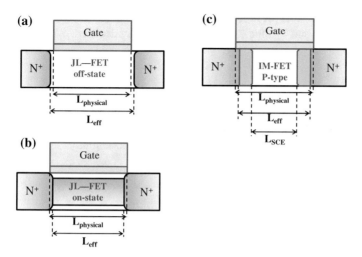

Fig. 6.2 Comparison among different L_{eff}. **a** The L_{eff} with the channel of JL—FET in off-state. **b** The L_{eff} with the channel of JL—FET in on-state. **c** The existence of junction when the channel of IM-FET in off-state will lead to smaller L_{SCE}, thus making it vulnerable to SCE

6.3   JL—FET Operating Mechanism

By compared to IM-FET, the difference is the dopant type and dopant concentration in the active layer as shown in Fig. 6.2. The source, drain, and channel of JL—FET are all based on the same doping and concentration. The comparison between different operating mechanisms of (a) inversion-mode (IM) "N$^+$PN$^+$" FET, (b) accumulation-mode (AC) "N$^+$NN$^+$" FET, (c) junctionless-mode (JL) "N$^+$N$^+$N$^+$" FET are as shown in Fig. 6.3.

(a) In n-type IM-FET, the depletion region will take place between flatband voltage (V_{fb}) and V_{th}. If it is greater than V_{th}, there will be a strong inversion on the surface of channel thus forming an inversion layer.

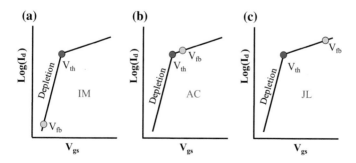

Fig. 6.3 Drain current as the function of V_{g}, **a** inversion-mode (IM), **b** accumulation-mode (AC), **c** junctionless-mode (JL) [7]

(b) In n-type AC-FET, it is fully depleted when V_g is lower than V_{th}. When V_g is between V_{th} and V_{fb}, the channel will be partially depleted. When V_g continues to rise, an accumulation layer will be formed on the surface of channel.

(c) In n-type JL—FET, source, drain and channel are all heavily doped around 10^{19}–10^{20} cm^{-3}. The channel is fully depleted when V_g is lower than V_{th}. The electron concentration in the channel will be increased along with increasing V_g. The channel concentration will approach maximum concentration when V_g reaches V_{th} with dopant concentration of N_d. When $V_g = V_{fb} > V_{th}$, the channel concentration is identical to source and drain concentration.

The electron concentration cross sections of n-type JL—FET are as shown in Fig. 6.4. The diagram of fixed $V_d = 50$ mV versus different V_g are as shown in Fig. 6.4a–d. These phenomena are obtained from the results of simulations based on Poisson Equation and the drift-diffusion model (Fig. 6.4).

(a) When V_g is lower than V_{th}, the n-type channel will be fully depleted, and the device is in OFF state.

(b) When V_g is equaled to V_{th}, the electron concentration of n-type silicon channel gradually becomes close to the dopant concentration while connecting drain and source, and free electrons are gradually generated in the channel along with the reduction of depletion region.

(c) When V_g is higher than V_{th}, the electron concentration of n-type silicon channel has become identical to the dopant concentration.

(d) When V_g reaches V_{fb}, the electron concentration at the junction of n-type silicon channel dopant concentration is identical to source and drain concentration. The JL—FET current conduction behavior can be regarded as a resistor.

The drain current of JL—FET is different from that of IM-FET. The former is more like a resistor relying on the drift current of majority carriers, and the latter relies on the drift current of minority inverse carriers. The equation of drain saturation current of general long-channel IM-FET (I_{dsat}) is shown as:

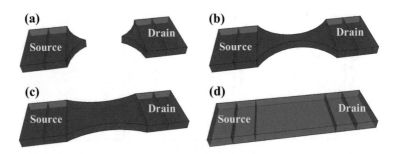

Fig. 6.4 Electron concentration cross sections of n-type JL—FET with fixed $V_d = 50$ mV. **a** $V_g < V_{th}$; **b** $V_g = V_{th}$; **c** $V_g > V_{th}$; **d** $V_g = V_{fb} \gg V_{th}$ [7]

$$I_{dsat} \approx \mu C_{ox} \frac{W}{L} (V_g - V_{th})^2 \tag{6.2}$$

where W is the channel width, L is the gate length, V_g is the gate voltage, and C_{ox} is the capacitor of oxide layer. On the other hand, JL—FET is a normally on device just like a resistor. When V_g is equaled to V_{fb}, the I_{dsat}:

$$I_{dsat} \approx q\mu N_d \frac{T_{si} W_{si}}{L} V_g \tag{6.3}$$

where T_{si} is the thickness of silicon channel, and N_d is the dopant concentration. It is shown in this equation that I_{dsat} is not related to C_{ox}.

The current conduction positions of IM-FET, AC-FET, and JL—FET are as shown in Fig. 6.5. In IM-FET when it is lower than V_{th}, the current conduction position is at the corner because of greater electric field here as shown in Fig. 6.5a. When it is higher than V_{th}, the current takes place in the channel along the edge of gate as shown in Fig. 6.5d; in AC-FET when it is lower than V_{th}, the sub-threshold current is formed in the center of channel as shown in Fig. 6.5b. When it is higher than V_{th}, it will result in the surface current similar to inversion-mode as shown in Fig. 6.5e; in the end, in JL—FET when it is lower than V_{th}, the sub-threshold current is also formed in the center of channel just like AC-FET as shown in Fig. 6.5c. When it is higher than V_{th}, the channel is partially depleted, and the main current is still contributed by the center of channel as shown in Fig. 6.5f. The conducting current is concentrated in the center, which is called body current. It is different to the conventional MOSFET as surface current underneath of gate oxide. The advantage of this body current of JL—FET can reduce the interface scattering effects in the channel.

Design and simulation of 3D JL—FET by using Synopsys Sentaurus TCAD 2014 version will be described in this chapter.

Fig. 6.5 Various current distributions of inversion-mode, accumulation-mode, and junctionless-mode FETs. The FETs in the state of "Below Threshold" are as shown in the upper half, and the FETs in the state of "Above Threshold" are as shown in the lower half

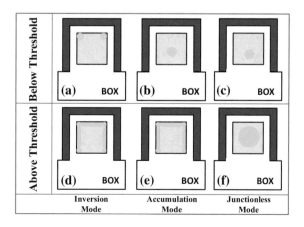

6.4   [Example 6.1] n-Type JL—FET with L_g = 10 nm

The fundamental structure of JL—FET is shown in Fig. 6.1. As compared to traditional IM-FET, the channel dopant is the same as source (S) and drain (D). Therefore, source, drain, and channel of Example 6.1 are all based on n-type dopants. Only the part of doping needs to be modified in devise_dvs.cmd is shown below, for the rest of codes please refer to Chap. 3 (Fig. 6.6).

1. SDE – devise_dvs.cmd
This is the standard example of **3D JL—FET.**
The symbol ; is the prompt character for program designer to take note such that it will not be executed by the computer

```
;--- example 6-1 JL-nFinFET--------------------;
;-------------- parameter --------------------;
(define nm 1e-3)
(define Fw 5)
(define Fh 5)
(define Lg @Lg@)
(define LSDC 15)
(define LSD 15)
(define Tox 1)
(define x1 LSDC)
(define x2 (+ x1 LSD))
(define x3 (+ x2 Lg))
(define x4 (+ x3 LSD))
(define x5 (+ x4 LSDC))
(define y1 Fw)
(define y2 (+ y1 Tox))
(define y3 (+ y2 10))
(define z1 Fh)
(define z2 (+ z1 Tox))
(define C_Doping 1.5e19)
(define SD_Doping 1.5e19)
(define SDC_Doping 1.5e19)
```

```
;------------------------------------- Doping -----------------------------------------;
;----- Channel -----;
(sdedr:define-constant-profile "dopedC" "ArsenicActiveConcentration" C_Doping )
(sdedr:define-constant-profile-region   "RegionC" "dopedC" "Channel" )
;----- Source -----;
(sdedr:define-constant-profile "dopedS" "ArsenicActiveConcentration" SD_Doping )
(sdedr:define-constant-profile-region   "RegionS" "dopedS" "Source" )
(sdedr:define-constant-profile "dopedSC" "ArsenicActiveConcentration" SD_Doping )
(sdedr:define-constant-profile-region   "RegionSC" "dopedSC" "SourceC" )
;----- Drain ------;
(sdedr:define-constant-profile "dopedD" "ArsenicActiveConcentration" SD_Doping )
(sdedr:define-constant-profile-region   "RegionD" "dopedD" "Drain" )
(sdedr:define-constant-profile "dopedDC" "ArsenicActiveConcentration" SD_Doping )
(sdedr:define-constant-profile-region   "RegionDC" "dopedDC" "DrainC" )
;------------------------------- END -----------------------------------;
```

The I_d–V_g curve of simulation result is shown in Fig. 6.7 with important parameters as shown in Fig. 6.8. From Figs. 6.7 and 6.8, it appears that the SS of L_g = 10 nm p-type JL—FET is around 73 mV/dec. with excellent switch characteristic. The structural channel mesh, the electron concentration distributions of 3D and 2D structures, electric field distributions, electric potential distributions, and the energy band diagrams along the channel direction are as shown in Figs. 6.9, 6.10, 6.11, 6.12, 6.13, 6.14, 6.15, 6.16 and 6.17 with the conditions of L_g = 10 nm, V_d = 1 V, and V_g = 1 V, respectively. The additional texts in Figs. 6.9, 6.10, 6.11, 6.12, 6.13, 6.14, 6.15, 6.16 and 6.17 are added via PowerPoint for better understanding by readers.

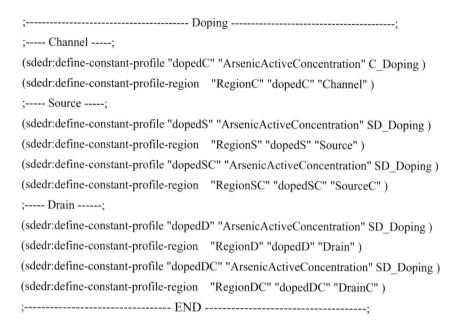

Fig. 6.6 Required simulation tools are shown in the workbench for n-type JL—FET

3D-nJLFET

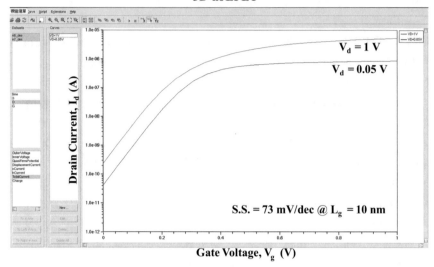

Fig. 6.7 I_d–V_g curve of 3D simulation of L_g = 10 nm n-type JL—FET (some descriptions are completed by PowerPoint after snapshot by Inspect)

	Vd	VT	gmax	ioff	isat	sslop	rout
1	0.05	--	0.2716159158256913	2.89950297455037e-06	4.30680166392608e-11	8.51392778288861e-07	0.0731581372312524	737709.2744496589
2	1	--	0.2079882496635378	9.207053959314701e-06	2.38733368764584e-10	5.0973106647299e-06	0.0725612540411235	131998.5236865284

Fig. 6.8 Electric property parameters of 3D simulation of L_g = 10 nm n-type JL—FET

3D-nJLFET Mesh

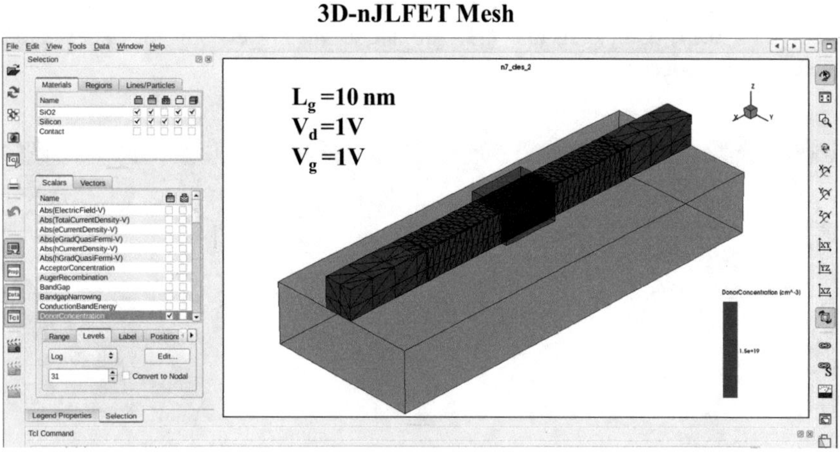

Fig. 6.9 Mesh diagram of 3D simulation of L_g = 10 nm n-type JL—FET

3D-nJLFET Mesh

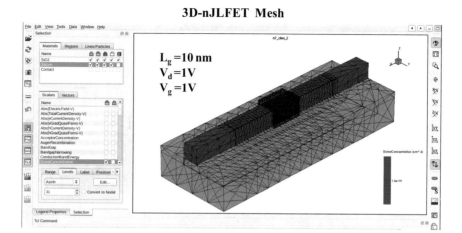

Fig. 6.10 Mesh diagram of 3D simulation of L_g = 10 nm n-type JL—FET (including Silicon dioxide)

3D-Electron concentration

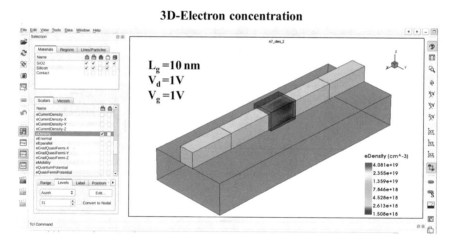

Fig. 6.11 Electron concentration distribution of 3D simulation of L_g = 10 nm n-type JL—FET

6.5 [Example 6.2] p-Type JL—FET with L_g = 10 nm

The fundamental structure of JL—FET is shown in Fig. 6.1. As compared to traditional IM-FET, the channel dopant is the same as source (S) and drain (D). Therefore, source, drain, and channel of Example 6.2 are all based on p-type dopants. Only the part of doping needs to be modified in devise_dvs.cmd is shown below, for the rest of codes please refer to Chap. 3 (Fig. 6.18).

2D-Electron concentration

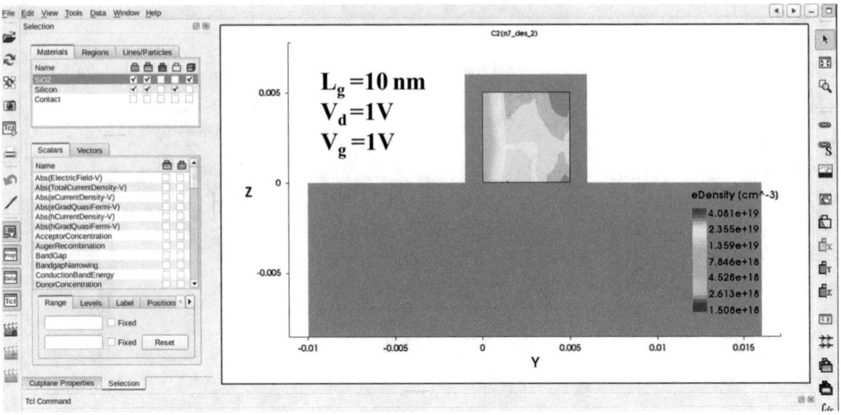

Fig. 6.12 2D Electron concentration distribution of 3D simulation of L_g = 10 nm n-type JL—FET is **body current**

3D-Electric Field

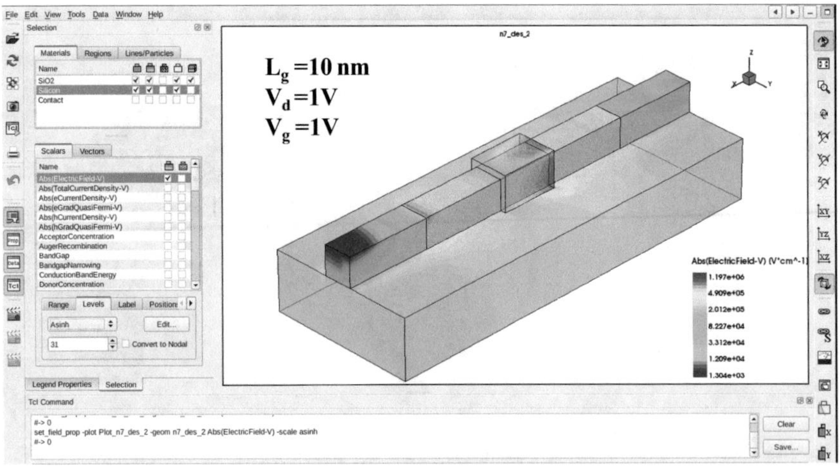

Fig. 6.13 Electric field distribution of 3D simulation of L_g = 10 nm n-type JL—FET

1. SDE – devise_dvs.cmd

This is the standard example of **3D JL—FET**.

The line of code following; **is the prompt character for program designer to take note such that it will not be executed by the computer**.

2D-Electric Field

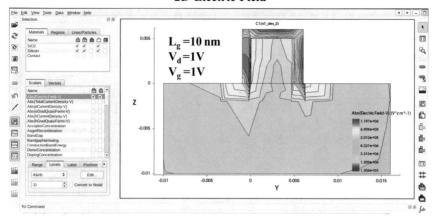

Fig. 6.14 2D electric field distribution of 3D simulation of L_g = 10 nm n-type JL—FET

3D-Electrostic Potential

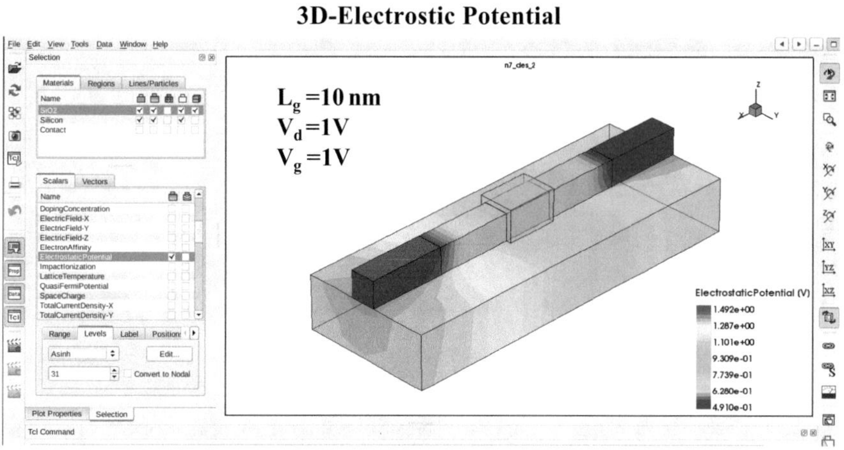

Fig. 6.15 Electric potential distribution of 3D simulation of L_g = 10 nm n-type JL—FET

2D-Electrostic Potential

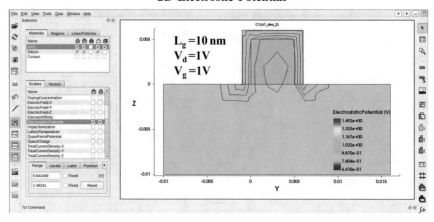

Fig. 6.16 2D electric potential distribution of simulation of 3D L_g = 10 nm n-type JL—FET

Band Diagram

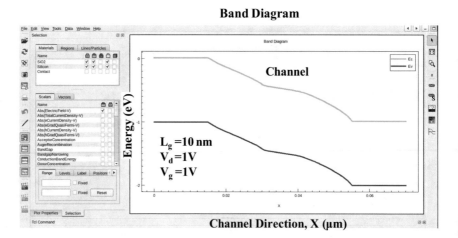

Fig. 6.17 Energy band diagram along the channel direction of simulation of 3D L_g = 10 nm n-type JL—FET in on-state

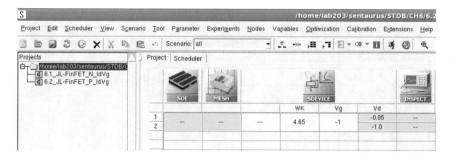

Fig. 6.18 Required simulation tools are shown in the workbench for p-type JL—FET

;------------------------------------- Doping --------------------------------------;
;----- Channel -----;
(sdedr:define-constant-profile "dopedC" "BoronActiveConcentration" C_Doping)
(sdedr:define-constant-profile-region "RegionC" "dopedC" "Channel")
;----- Source -----;
(sdedr:define-constant-profile "dopedS" "BoronActiveConcentration" SD_Doping)
(sdedr:define-constant-profile-region "RegionS" "dopedS" "Source")
(sdedr:define-constant-profile "dopedSC" "BoronActiveConcentration" SD_Doping)
(sdedr:define-constant-profile-region "RegionSC" "dopedSC" "SourceC")
;----- Drain ------;
(sdedr:define-constant-profile "dopedD" "BoronActiveConcentration" SD_Doping)
(sdedr:define-constant-profile-region "RegionD" "dopedD" "Drain")
(sdedr:define-constant-profile "dopedDC" "BoronActiveConcentration" SD_Doping)
(sdedr:define-constant-profile-region "RegionDC" "dopedDC" "DrainC")
--- END ---

The I_d–V_g curve of simulation result is as shown in Fig. 6.19 with important parameters as shown in Fig. 6.20. From Figs. 6.19 and 6.20, it appears that the SS of L_g = 10 nm p-type JL—FET is around 73 mV/dec. with good electric properties.

The structural channel mesh, the electron concentration distributions of 3D and 2D structures, electric field distributions, electric potential distributions, and the energy band diagrams along the channel direction are as shown in Figs. 6.21, 6.22, 6.23, 6.24, 6.25, 6.26, 6.27, 6.28 and 6.29 with the conditions of L_g = 10 nm, V_d = −1 V, and V_g = −1 V, respectively. The additional texts in Figs. 6.21, 6.22, 6.23, 6.24, 6.25, 6.26, 6.27, 6.28 and 6.29 are added via PowerPoint for better understanding by readers.

3D-pJLFET

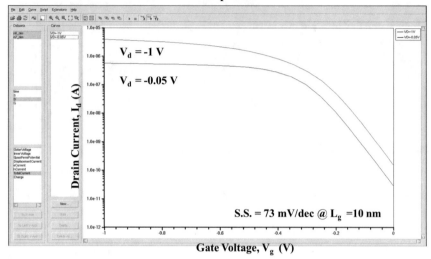

Fig. 6.19 I_d–V_g curve of simulation of 3D L_g = 10 nm p-type JL—FET

	Vd		VT	gmax	ioff	isat	sslop	rout
1	-0.05	--	-0.302320581544836	-8.908937021123012e-10	2.88438987393031e-11	5.76743051775071e-07	0.0745504842798749	-497.789039841590
2	-1	--	-0.228442984604629	-4.849648228023753e-09	1.53965380916126e-10	3.89849227294142e-06	0.0731030288363683	-142.714213303438

Fig. 6.20 Electric property parameters of 3D simulation of L_g = 10 nm p-type JL—FET

3D-pJLFET Mesh

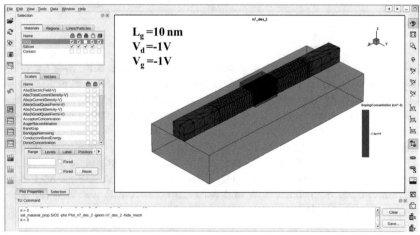

Fig. 6.21 Mesh diagram of 3D simulation of L_g = 10 nm p-type JL—FET

3D-nJLFET Mesh

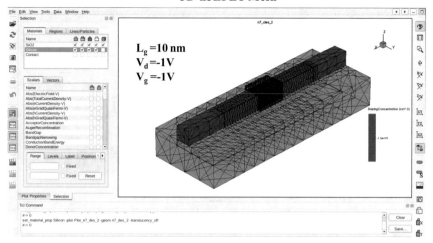

Fig. 6.22 Mesh diagram of 3D simulation of L_g = 10 nm p-type JL—FET (including Silicon dioxide)

3D-Hole concentration

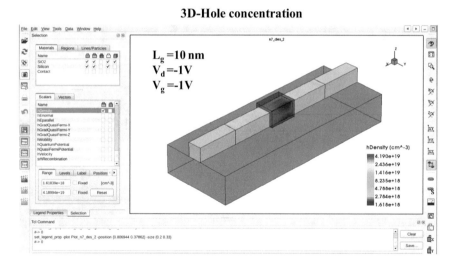

Fig. 6.23 Hole concentration distribution of 3D simulation of L_g = 10 nm p-type JL—FET

2D-Hole concentration

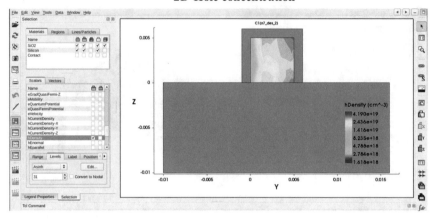

Fig. 6.24 2D hole concentration distribution of 3D simulation of L_g = 10 nm p-type JL—FET is the **body current**

3D-Electric Field

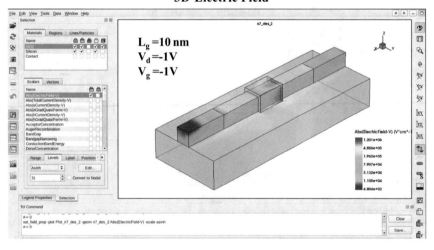

Fig. 6.25 Electric field distribution of 3D simulation of L_g = 10 nm p-type JL—FET

2D-Electric Field

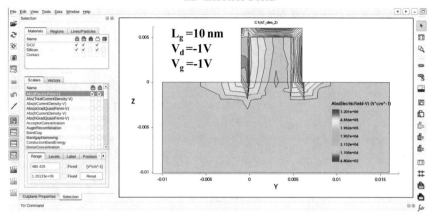

Fig. 6.26 2D electric field distribution of 3D simulation of L_g = 10 nm p-type JL—FET

3D-Electrostic Potential

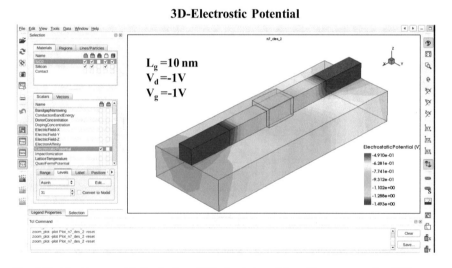

Fig. 6.27 Electric potential distribution of 3D simulation of L_g = 10 nm p-type JL—FET

2D-Electrostic Potential

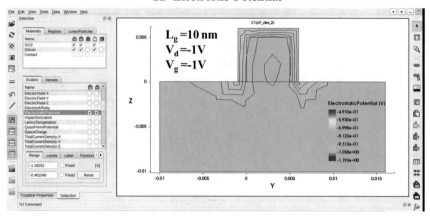

Fig. 6.28 2D electric potential distribution of 3D simulation of L_g = 10 nm p-type JL—FET

Band Diagram

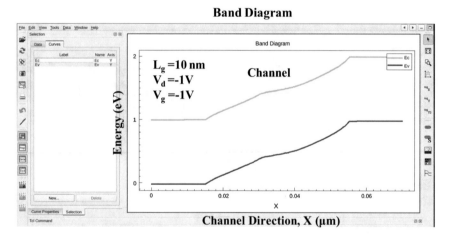

Fig. 6.29 Energy band diagram along the channel direction of simulation of 3D L_g = 10 nm p-type JL—FET in on-state

In summary, the standard TCAD simulation example of 3D JL—FET has been provided in this chapter. JL—FET will not suffer from the issues of PN junction and dopant concentration gradient, and it is based on a relatively simple process with low thermal budget, making it feasible for gate first process. Therefore, it is filled with potential for the application of future sub-10 nm technology node. The JL—FET operation principle and simulation method can be used in high mobility materials, including Graphene, MoS2 et al. in future JL—FETs.

References

1. H.B. Chen, Y.C. Wu, C.Y. Chang, M.H. Han, N.H. Lu, Y.C. Cheng, Performance of GAA poly-Si nanosheet (2 nm) channel of junctionless transistors with ideal subthreshold slope. *VLSI Technology Symposium*, p T232 (2013)
2. Y.C. Cheng, H.B. Chen, C.S. Shao, J.J. Su, Y.C. Wu, C.Y. Chang, T.C. Chang, Performance enhancement of a novel P-type junctionless transistor using a hybrid Poly-Si fin channel. *Technical Digest of IEDM*, 26.27.21 (2014)
3. M.S. Yeh, Y.C. Wu, M.H. Wu, Y.R. Jhan, M.H. Chung, M.F. Hung, High performance ultra-thin body (2.4 nm) poly-Si junctionless thin film transistors with a trench structure. *Technical Digest of IEDM*, 26.26.21 (2014)
4. S. Migita, Y. Morita, M. Masahara, H. Ota, Electrical performances of junctionless-FETs at the scaling limit (L_{ch} = 3 nm). *Technical Digest of IEDM,* 8.6.1 (2012)
5. J.P. Colinge, I. Ferain, G. Fagas, S. Das, P. Razavi, R. Ya, Influence of channel material properties on performance of nanowire transistors. J. Appl. Phys. **111**, 124509 (2012)
6. J.P. Colinge, I. Ferain, A. Kranti, C.W. Lee, N.D. Akhavan, P. Razavi, R. Ya, R. Yu, Junctionless nanowire transistor: complementary metal-oxide-semiconductor without junctions. Sci. Adv. Mater. **3**, 477 (2011)
7. J.P. Colinge, C.W. Lee, N. Dehdashti Akhavan, R. Yan, I. Ferain, P. Razavi, A. Kranti, R. Yu, *Semiconductor-On-Insulator Materials for Nanoelectronics Applications. Engineering Materials*, Springer, Berlin (2011)

Chapter 7
Steep Slope Tunnel FET Simulation

7.1 Problems Facing Conventional MOSFET

The simplest way for increasing the transistor density in the wafer is to reduce the feature size of transistor. During scaling down of feature size by **Moore's law**, the supply voltage V_{dd} must also be reduced in accordance with the principle of constant electrical field scaling rule. And the threshold voltage V_{th} must also be reduced along with the reduction of V_{dd} in order to maintain the consistent **overdrive voltage ($V_{ov} = V_{dd} - V_{th}$)** and I_{on} as shown in Fig. 7.1.

For the fixed $V_{dd} = 1$ V, by reduction of V_{th}, the high-performance (high V_{ov} and high I_d) device will also lead to high leakage current (I_{off}). On the other hand, by increasing V_{th}, the I_{off} reduce will lead to low I_d and device performance. The phenomenon comes from the subthreshold slope **(SS) > 60 mV/dec. (SS = kT/ q × ln10 ~60 mV/dec.)** and will result in significant increase of off-state current level. The correlations among V_{ov}, SS, V_{th}, I_{off}, and I_{on} are as shown in Fig. 7.1.

The calculation of IC energy consumption is very important, especially when the requirement of current portable products, which need low power consumption. The total consumed energy can be divided into dynamic energy consumption ($E_{dynamic}$) and static energy consumption ($E_{leakage}$), and the relationship equation between total consumed energy (J) and V_{dd} is as shown below [1]:

$$
\begin{aligned}
E_{total} = E_{dynamic} + E_{leakage} &= \alpha L_d C V_{dd}^2 + L_d I_{off} V_{dd} \tau_{delay} \\
&\approx \alpha L_d C V_{dd}^2 + L_d C V_{dd}^2 \frac{I_{off}}{I_{on}} = L_d C V_{dd}^2 \left(\alpha + \frac{I_{off}}{I_{on}} \right) \\
&\approx L_d C V_{dd}^2 \left(\alpha + 10^{\frac{-V_{dd}}{SS}} \right)
\end{aligned}
\tag{7.1}
$$

$$
\text{wherer } \tau_{delay} = C V_{dd}/I_{on}
$$

© Springer Nature Singapore Pte Ltd. 2018
Y.-C. Wu and Y.-R. Jhan, *3D TCAD Simulation for CMOS Nanoelectronic Devices*,
DOI 10.1007/978-981-10-3066-6_7

Fig. 7.1 I_d–V_g characteristic curve of conventional nMOSFET with drain current I_d and gate voltage (V_g) with the identical SS value and different V_{th}

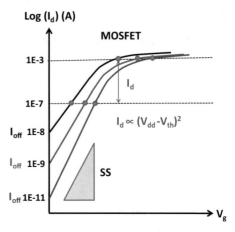

where L_d is logic depth, C is switched capacitance, τ_{delay} is delay time, and α is logic activity factor (usually is 0.01). The operating frequency (f) and delay time τ_{delay} can be expressed as:

$$f = \frac{1}{L_d \tau_{delay}} \tag{7.2}$$

$$\tau_{delay} = \frac{CV_{dd}}{I_{on}} \tag{7.3}$$

It is clearly shown in Eq. (7.1) and Fig. 7.1 that the transistor with low standby power consumption can be achieved by working on small SS and low V_{dd}. V_{dd} will depend on the operating voltage $V_{ov} = (V_{dd} - V_{th})$, and the excessive reduction of this voltage will result in another issue of low I_{on}. Therefore, we can reduce the off-state current I_{off} by another method of reducing SS value.

7.2   Operating Mechanism of Tunnel FET (TFET)

It is shown in Fig. 7.1 that I_{off} is limited by V_{th} and subthreshold slope (SS). The SS physical limit is 60 mV/dec. If SS < 60 mV/dec., I_{off} can be greatly reduced in Fig. 7.2. The SS of conventional MOSFET is resulted from thermal diffusion current, such that it is bound to be >60 mV/dec [1]. The current of tunnel FET (TFET) is resulted from tunneling current rather than thermal diffusion current, such that the SS of TFET can break the physical limit of 60 mV/dec. In recent years, there have been many international research groups dedicated to study on innovative TFET structures and materials to break SS of 60 mV/dec, and also increasing the ON current [2–6].

Figure 7.3 shows TFET is a perfect choice for **ultra-low-power device with low V_{th} and low V_{dd}**, which is urgently needed for Internet of Thing (IoT) ultra-low-power applications as shown in Fig. 7.4.

Fig. 7.2 I_d–V_g characteristic curve of devices with different SS

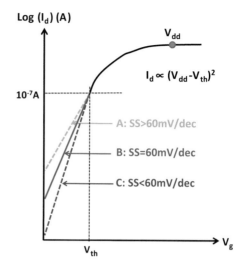

Fig. 7.3 Comparison among subthreshold slopes (SS) of various semiconductor devices [1]

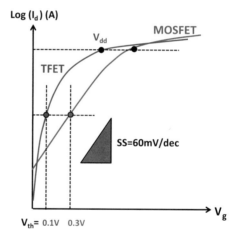

The fundamental structure of TFET is as shown in Fig. 7.5, and the basic operating mechanism of TFET based on the principle of "Gate-controlled reverse PIN diode", which is shown in Fig. 7.6. Unlike conventional n-channel MOSFET, the drain of n-channel TFET is based on n-type doping, and the source is based on p-type doping. For n-channel TFET, the electrons tunnel from p-type doping source to n-type doping drain. The TFET operation is shown in Fig. 7.6.

In the off-state, electrons in drain and holes in source will both face rather high energy barrier thus preventing conduction of both electrons and holes, and the leakage current I_{off} is also fairly low. In the on-state, the gate voltage is bigger than threshold voltage (V_{th}), and a significant band bending will occur in the depletion region in the area of source close to the channel, and the width of depletion region will be gradually reduced along with the increasing gate voltage (V_g). Under such

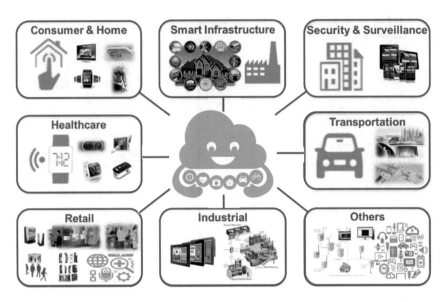

Fig. 7.4 Coming era of IoT and the applications of various semiconductor devices requiring ultra-low-voltage operations

Fig. 7.5 Fundamental structure of n-type tunnel (TFET) based on Gate-controlled reverse PIN diode

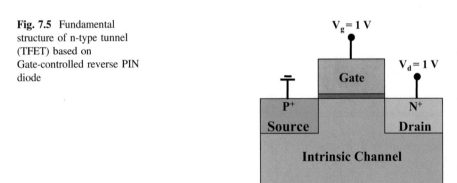

conditions, electrons in the valence band of source will have the chance to tunnel through the narrow depletion region ($\sim$nm) and enter the conduction band of channel, and this is the unique tunneling current conduction mechanism of TFET. The conduction current is called tunneling current which is proportional to the tunneling probability of WKB approximation is based on the equation as shown below:

$$I_{\text{on}} \propto T_{\text{wkb}} \approx \exp\left(-\frac{4\lambda\sqrt{2m^*}E_g^{3/2}}{3q\hbar(E_g+\Delta\Phi)}\right) \quad (7.4)$$

where λ is the screening tunneling length, m^* is the effective mass of carrier, E_g is energy band gap, and $\Delta\Phi$ is the energy difference (tunneling window) between the

Fig. 7.6 Energy band diagram of tunnel FET (TFET)

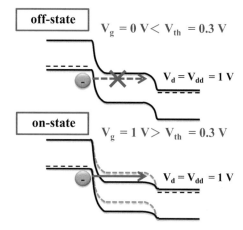

of valence band (Ev) of source and the energy of conduction band (Ec) of channel (Fig. 7.5).

The conduction current of TFET is also related to the difference between the values of Fermi-Dirac distribution functions of channel and source and tunneling probability as shown in the equation below:

$$I_{on} = I_{s \to ch} - I_{ch \to s}$$
$$= C_1 \int \{D_s(E)f_s(E)T_{wkb}D_{ch}(E)[1 - f_{ch}(E)] - D_{ch}(E)f_{ch}(E)T_{wkb}D_s(E)[1 - f_s(E)]\}dE$$
$$= C_1 \int \{D_s(E)D_{ch}(E)T_{wkb}[f_s(E) - f_{ch}(E)]\}dE$$

$$(7.5)$$

where D_{ch} and D_s are the density of states of channel and source, respectively, and $f_{ch}(E)$ and $f_s(E)$ are the Fermi-Dirac functions of channel and source.

7.3 Example 7.1 (Design and Simulation of 3D n-Type TFET)

The following 3D nTFET three main program code files are based on Synopsys Sentaurus TCAD 2014 version.

In the introduction of TFET fundamental structure in Fig. 7.5, the channel is intrinsic without any doping. However, the channel in Example 7.1 is based on a slight amount of n-type doping for the purpose of adjusting threshold voltage (V_{th}).

1. **SDE—devise_dvs.cmd**

The method for establishing SDE of 3D nTFET is similar to Chap. 3 such that the code will not be introduced in details here. For detailed information please refer to Chap. 3. Here, we only introduce the **SDE tool codes: devise_dvs.cmd**, especially note the doping section.

```
;---------------------------------- parameter ----------------------------------;
(define nm 1e-3)
(define Fw 5)
(define Fh 5)
(define Lg @Lg@)
(define LSDC 15)
(define LSD 15)
(define Tox 2)
(define x1 LSDC)
(define x2 (+ x1 LSD))
(define x3 (+ x2 Lg))
(define x4 (+ x3 LSD))
(define x5 (+ x4 LSDC))
(define y1 Fw)
(define y2 (+ y1 Tox))
(define y3 (+ y2 10))
(define z1 Fh)
(define z2 (+ z1 Tox))
(define m1 (/ Lg 10))
(define m2 (/ Lg 20))
(define C_Doping 1e17)
(define S_Doping 1e20)
(define SC_Doping 1e20)
(define D_Doping 1e19)
(define DC_Doping 1e19)
;(define B_Doping 5e18)


;---------------------------------- Structure ----------------------------------;
```

"ABA"
;--- Source contact and Source ---;
(sdegeo:create-cuboid (position 0 0 0) (position x1 y1 z1) "Silicon" "SourceC")
(sdegeo:create-cuboid (position x1 0 0) (position x2 y1 z1) "Silicon" "Source")
;--- Gate oxide ---;
(sdegeo:create-cuboid (position x2 (- Tox) 0) (position x3 y2 z2) "SiO2"
"Gateoxide")
;--- Channel ---;
(sdegeo:create-cuboid (position x2 0 0) (position x3 y1 z1) "Silicon" "Channel")
;--- Drain contact and Drain---;
(sdegeo:create-cuboid (position x3 0 0) (position x4 y1 z1) "Silicon" "Drain")
(sdegeo:create-cuboid (position x4 0 0) (position x5 y1 z1) "Silicon" "DrainC")
;--- Buried oxide ---;
(sdegeo:create-cuboid (position 0 (- 10) (- 10)) (position x5 y3 0) "SiO2" "Box")
;-------------------------------------- Contact --------------------------------------;
;----- Source -----;
(sdegeo:define-contact-set "S" 4.0 (color:rgb 1.0 0.0 0.0) "##")
(sdegeo:set-current-contact-set "S")
(sdegeo:set-contact-faces (find-face-id (position 1 1 z1)))
;----- Drain -----;
(sdegeo:define-contact-set "D" 4.0 (color:rgb 1.0 0.0 0.0) "##")
(sdegeo:set-current-contact-set "D")
(sdegeo:set-contact-faces (find-face-id (position (+ x4 1) 1 z1)))
;----- Front Gate -----;
(sdegeo:define-contact-set "G" 4.0 (color:rgb 1.0 0.0 0.0) "||")
(sdegeo:set-current-contact-set "G")
(sdegeo:set-contact-faces (find-face-id (position (+ x2 1) (- Tox) 1)))
;----- Top Gate -----;
(sdegeo:define-contact-set "G" 4.0 (color:rgb 1.0 0.0 0.0) "||")
(sdegeo:set-current-contact-set "G")
(sdegeo:set-contact-faces (find-face-id (position (+ x2 1) 1 z2)))
;----- Back Gate -----;
(sdegeo:define-contact-set "G" 4.0 (color:rgb 1.0 0.0 0.0) "||")
(sdegeo:set-current-contact-set "G")
(sdegeo:set-contact-faces (find-face-id (position (+ x2 1) y2 1)))
;-------------------------------------- **Doping** --------------------------------------;

```
;----- Channel -----;
(sdedr:define-constant-profile "dopedC" "ArsenicActiveConcentration" C_Doping )
(sdedr:define-constant-profile-region  "RegionC" "dopedC" "Channel" )
;----- Source -----;
(sdedr:define-constant-profile "dopedS" "BoronActiveConcentration" S_Doping )
(sdedr:define-constant-profile-region  "RegionS" "dopedS" "Source" )
(sdedr:define-constant-profile "dopedSC" "BoronActiveConcentration" S_Doping )
(sdedr:define-constant-profile-region  "RegionSC" "dopedSC" "SourceC" )
;----- Drain ------;
(sdedr:define-constant-profile "dopedD" "ArsenicActiveConcentration" D_Doping )
(sdedr:define-constant-profile-region  "RegionD" "dopedD" "Drain" )
(sdedr:define-constant-profile "dopedDC" "ArsenicActiveConcentration" D_Doping )
(sdedr:define-constant-profile-region  "RegionDC" "dopedDC" "DrainC" )

;---------------------------------------- Mesh ----------------------------------------------;
;--- AllMesh ---;
(sdedr:define-refinement-size "Cha_Mesh" 20 20 20 10 10 10)
(sdedr:define-refinement-material "channel_RF" "Cha_Mesh" "Silicon" )
;--- ChannelMesh ---;
(sdedr:define-refinement-window "multiboxChannel" "Cuboid"
(position x1 0 0)   (position x4 y1 z1))
(sdedr:define-multibox-size "multiboxSizeChannel"  m1 m1 m1 m2 m2 m2)
(sdedr:define-multibox-placement "multiboxPlacementChannel"
"multiboxSizeChannel" "multiboxChannel")
(sdedr:define-refinement-function "multiboxPlacementChannel"
"DopingConcentration" "MaxTransDiff" 1)

;----------    Save BND and CMD and rescale to nm ---------;
(sde:assign-material-and-region-names (get-body-list) )
(sdeio:save-tdr-bnd (get-body-list) "n@node@_nm.tdr")
(sdedr:write-scaled-cmd-file "n@node@_msh.cmd" nm)
(define sde:scale-tdr-bnd
 (lambda (tdrin sf tdrout)
    (sde:clear)
    (sdegeo:set-default-boolean "XX")
    (sdeio:read-tdr-bnd tdrin)
    (entity:scale (get-body-list) sf)
    (sdeio:save-tdr-bnd (get-body-list) tdrout)
    )
 )
(sde:scale-tdr-bnd "n@node@_nm.tdr" nm "n@node@_bnd.tdr")
;---------------------    END    ----------------------------;
```

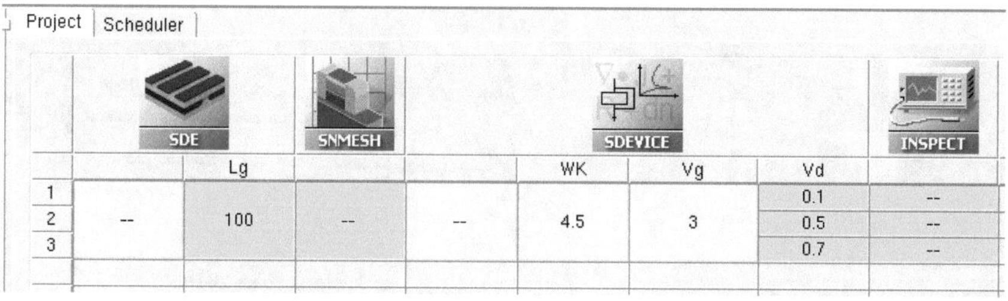

		Lg			WK	Vg	Vd	
1							0.1	--
2	--	100	--	--	4.5	3	0.5	--
3							0.7	--

Fig. 7.7 Required simulation tools are shown in the workbench for nTFET

2. **SDEVICE—dessis_des.cmd**

   The method for establishing SDEVICE of 3D nTFET is similar to s in Chap. 3 such that the code will not be introduced in details here. For detailed information please refer to Chap. 3. The current of TFET is tunneling current, such that the physical model of tunneling **Band2Band(E2)** must be added in **dessis_des. cmd**. For the physical details please refer to SDEVICE manual of Sentaurus TCAD 2014 version.

   ```
   ..............................
   Physics{
           Mobility( DopingDep HighFieldsat Enormal )
           EffectiveIntrinsicDensity( OldSlotboom BandGapNarrowing
   (BennettWilson ) )
           Recombination( SRH(DopingDependence) Auger Band2Band(E2) )
           *ComputeIonizationIntegrals(WriteAll)
           eQuantumPotential
           hQuantumPotential }
   ..........................
   ```

3. **INSPECT—inspect_inc.cmd**

   The method for establishing INSPECT of 3D nTFET is the same as in Chap. 3 such that the code will not be introduced in details here. For detailed information please refer to Chap. 3.

   The I_d–V_g curve of simulation result is as shown in Fig. 7.8, and the important parameters are as shown in Fig. 7.9. Sometime, the point hopping occurs to IV characteristic in the subthreshold region; the SS extracted by Inspect can no longer serve as the reference. For obtaining precise SS, readers can obtain the text output

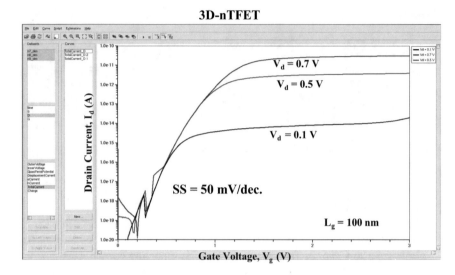

Fig. 7.8 I_d–V_g curve of simulation of 3D n-type TFET

	Vd		gmax	ioff	isat	sslop	rout
1	0.1	--	334544714823479e-	-2.50030463920446e-16	1.97355574899237e-14	1.043761982609599e-05	-267533287601009.55
2	0.5	--	83961873016234e-	-3.42754756919477e-18	3.9184544209425e-12	0.000111765418720418	-15876782314012.4
3	0.7	--	22008487426658e-	-4.87951536893058e-17	3.06795079102809e-11	0.000106812235106933	-1412268310784.859

Fig. 7.9 Electric property parameters of simulation of 3D n-type TFET

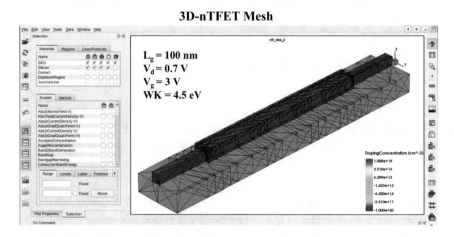

Fig. 7.10 Mesh diagram of the simulation of 3D n-type TFET (gated reverse PIN diode)

3D-Electron concentration

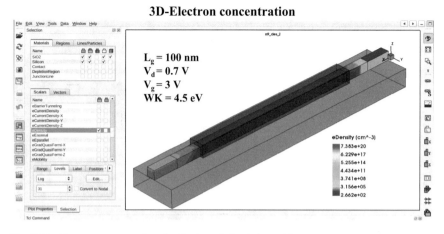

Fig. 7.11 Electron concentration distribution of simulation of 3D n-type TFET

2D-Electron concentration

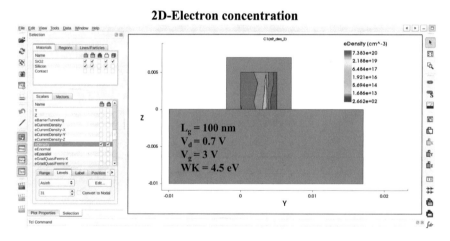

Fig. 7.12 2D electron concentration distribution of simulation of 3D n-type TFET

of IV characteristic for self-analysis and calculation. The SS in Fig. 7.8 is around 50 mV/dec. The structural channel mesh, the electron concentration distributions of 3D and 2D structures, electric field distributions, electric potential distributions, and the energy band diagrams along the channel direction are as shown in Figs. 7.10, 7.11, 7.12, 7.13, 7.14, 7.15, 7.16 and 7.17 with the conditions of $L_g = 100$ nm, $V_d = 0.7$ V, and $V_g = 3$ V, respectively. The additional texts in Figs. 7.10, 7.11, 7.12, 7.13, 7.14, 7.15, 7.16 and 7.17 are added via PowerPoint for better understanding by readers (Fig. 7.18).

3D-Electric Field

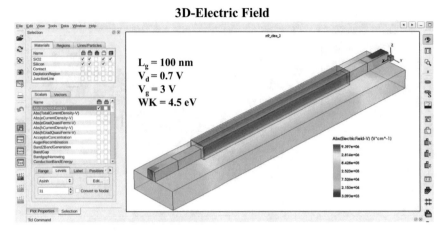

Fig. 7.13 Electric field distribution of simulation of 3D n-type TFET

2D-Electric Field

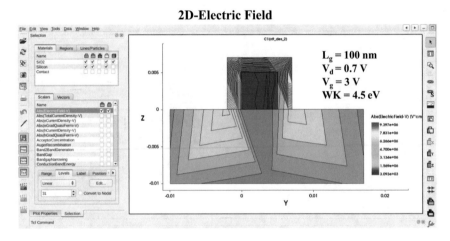

Fig. 7.14 2D electric field distribution of simulation of 3D n-type TFET

7.4   Example 7.2 (3D n-Type TFET of Different Drain Doping Concentrations)

The following three main program code files are based on Synopsys Sentaurus TCAD 2014 version. The **drain doping concentration D_Doping is set as variable**.

TFET is the transistor which can be subjected to bipolar operation, which means it can be turned on by either positive or negative bias. This phenomenon will cause

3D-Electrostatic Potential

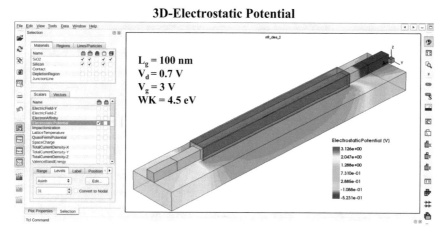

Fig. 7.15 Electric potential distribution of simulation of 3D n-type TFET

2D-Electrostatic Potential

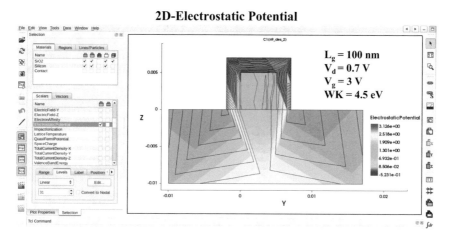

Fig. 7.16 2D electric potential distribution of simulation of 3D n-type TFET

the large off-state current (I_{off}) and it cannot be used for IC. This problem can be solved by reducing the dopant concentration of drain. The impact of reduced drain concentration of n-type TFET demonstrates in Example 7.2.

1. **SDE—devise_dvs.cmd**

   The method for establishing SDE of 3D nTFET is similar to Chap. 3. The only special part is about the doping. Therefore, only the program code of doping is introduced here. For the rest of codes please refer to Chap. 3.

Band Diagram

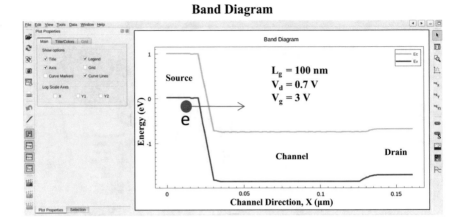

Fig. 7.17 Energy band diagram along the channel direction of simulation of 3D n-type TFET in on-state. The VB of source overlaps to CB of channel, and electrons will tunnel from source to drain through channel

| | SDE | | SNMESH | | SDEVICE | | | INSPECT |
	Lg	D_Doping			WK	Vg	Vd	
1							0.1	--
2		1e18	--	--	4.5	3	0.5	--
3							0.7	--
4							0.1	--
5		5e18	--	--	4.5	3	0.5	--
6							0.7	--
7							0.1	--
8	100	1e19	--	--	4.5	3	0.5	--
9							0.7	--
10							0.1	--
11		5e19	--	--	4.5	3	0.5	--
12							0.7	--
13							0.1	--
14		1e20	--	--	4.5	3	0.5	--
15							0.7	--

Fig. 7.18 Required simulation tools are shown in the workbench for nTFET with different drain doping concentrations

```
;-------------------------------------- Doping --------------------------------------;
;----- Channel -----;
(sdedr:define-constant-profile "dopedC" "ArsenicActiveConcentration" C_Doping )
(sdedr:define-constant-profile-region  "RegionC" "dopedC" "Channel" )
;----- Source -----;
(sdedr:define-constant-profile "dopedS" "BoronActiveConcentration" S_Doping )
(sdedr:define-constant-profile-region  "RegionS" "dopedS" "Source" )
(sdedr:define-constant-profile "dopedSC" "BoronActiveConcentration" S_Doping )
(sdedr:define-constant-profile-region  "RegionSC" "dopedSC" "SourceC" )
;----- Drain ------;
(sdedr:define-constant-profile "dopedD" "ArsenicActiveConcentration" D_Doping )
(sdedr:define-constant-profile-region  "RegionD"  "dopedD" "Drain" )
(sdedr:define-constant-profile "dopedDC" "ArsenicActiveConcentration" D_Doping )
(sdedr:define-constant-profile-region  "RegionDC" "dopedDC" "DrainC" )
;----------------------------------------------------------------------------------------;
```

It is indicated in the code that source is doped by boron and Drain is doped by arsenic. Unlike the aforementioned introduction, the arsenic of lower concentration is used as the dopant for the channel for the purpose of adjusting V_{th}

2. **SDEVICE—dessis_des.cmd**
   The method for establishing SDEVICE of 3D nTFET is the same as in Chap. 3, so the codes will not be introduced in details here. Readers can refer to Chap. 3 for detailed information.

3. **INSPECT—inspect_inc.cmd**
   The code of this part is completely identical to Example 7.1, so readers can directly refer to the code of Example 7.1.

The $I_{\text{d}}–V_{\text{g}}$ curve of simulation result of Example 7.2 is as shown in Fig. 7.19, with important parameters as shown in Fig. 7.20. It is indicated in Fig. 7.19 that the bipolarity of TFET can be effectively inhibited by reducing the dopant concentration of drain. In this simulation results, the I_{off} can be reduced by decreasing dopant concentration of drain. The drain doping 1E18 cm^{-3} shows lowest I_{off} than others. It can be explained that the lower drain doping has less energy band bending to prevent the leakage current.

3D-nTFET with Different Concentration

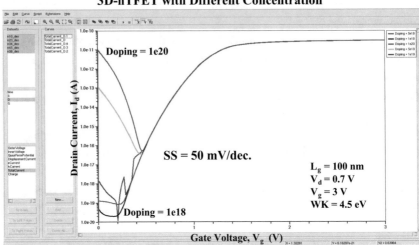

Fig. 7.19 I_d–V_g curves of simulation of 3D n-type TFET with different drain dopant concentrations

	Vd		gmax	ioff	isat	sslop	rout
1	0.1	--	)49992652715592e-	-4.26177203293538e-16	1.97357941871557e-14	1.04414392746969e-05	-54856936134.78269
2	0.5	--	909280878341169e-	-5.28803647186137e-19	3.91853005610195e-12	1.041828196082225e-05	-23941298609933.64
3	0.7	--	15578899100872e-	-2.3014565150108e-17	3.06799706868228e-11	1.040460607415789e-05	-1241031101551884
4	0.1	--	451437028407683e-	-2.53540833121077e-16	1.9735591646631e-14	1.045259171502318e-05	-1181748716501.113
5	0.5	--	88394375027677e-	-5.17256522733038e-16	3.9184692748669 1e-12	1.039552082756731e-05	-12204564665.91805
6	0.7	--	166720833104039e-	8.80763996250153e-20	3.06796054345289e-11	0.073619485165816	-2620088288240164
7	0.1	--	334544718823479e-	-2.50030463920446e-16	1.97355575404662e-14	1.043761982609599e-05	-26753328760109.55
8	0.5	--	339618730845575e-	-3.42754756919477e-18	3.9184544424 8412e-12	0.000111765418720418 8	-15876782314012.4
9	0.7	--	122008486926676e-	-4.87951628092964e-17	3.06795079099831e-11	0.000106812235106933	-1412279475905.992
10	0.1	--	391000202480928e-	-2.34067828879641e-20	1.97354949730045e-14	4.729156167079517e-05	-414838256151385.1
11	0.5	--	87343208016064 8e-	1.18358259685009e-18	3.91843190884422e-12	0.0767655251368 6678	-11727014678.05261
12	0.7	--	196708453761589e-	3.54140253373422e-17	3.06793823370024e-11	0.0965905697883 0172	-633182778.4131511
13	0.1	--	391019712764759e-	-3.11077620903149e-20	1.97354487081566e-14	1.042595883518921e-05	-113479660457896
14	0.5	--	)52360808085219e-	8.39995526947715e-19	3.91842716480476e-12	0.0725069806428 1731	-133296012.5431249
15	0.7	--	184869895348462e-	5.32410873765499e-17	3.06793629530829e-11	0.09622418878315821	-7346361.829692629

Fig. 7.20 Electric property parameters of simulation of 3D n-type TFET with different drain dopant concentrations

7.5 Example 7.3 (3D n-Type TFET with Asymmetrical Gate)

7.5.1 Descriptions of Motivation and Principle

This part describes an asymmetric-gate tunnel field effect transistor (AG-TFET) with a gate-all-around (GAA) structure in the source and a planar structure in the drain [7]. It has a low off-state current (6.55×10^{-16} A/μm) and a high on-state current (2.47×10^{-5} A/μm), because the screening length (λ) of a GAA nanowire (NW) structure is half that of the planar structure. Simulations reveal that a sub-threshold swing (*SS*) as low as 42 mV/dec, and an on/off current ratio as higher as 10^{10} is realized. The AG-TFET is easily fabricated as an actual device by simply changing the layout of gate in a general TFET fabrication.

1. **SDE—devise_dvs.cmd**
   The method for establishing SDE of AG-TFET is the same as in Chap. 3, so the codes will not be introduced in details here. Readers can refer to Chap. 3 for detailed information.

2. **SDEVICE—dessis_des.cmd**
   The method for establishing SDEVICE of AG-TFET is the same as in Chap. 3, so the codes will not be introduced in details here. Readers can refer to Chap. 3 for detailed information. The current of TFET is tunneling current, so the tun-neling physical model must be added in the physics part with the manual as the reference for details.

3. **INSPECT—inspect_inc.cmd**
   The method for establishing INSPECT of AG-TFET is the same as in Chap. 3, so the codes will not be introduced in details here. Readers can refer to Chap. 3 for detailed information.

Figure 7.21 displays the architecture of the AG-TFET and the parameters of its simulation. The total channel length is 20 nm. The left half (10 nm) of gate of AG-TFET is controlled by the surrounding gate with a square cross section, and the

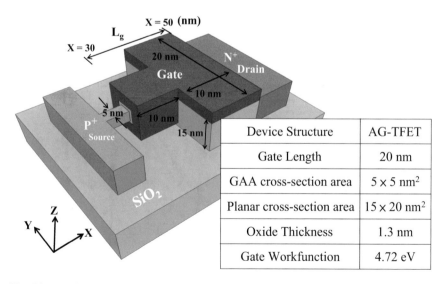

Device Structure	AG-TFET
Gate Length	20 nm
GAA cross-section area	5×5 nm^2
Planar cross-section area	15×20 nm^2
Oxide Thickness	1.3 nm
Gate Workfunction	4.72 eV

Fig. 7.21 Device structure and important simulation parameters of n-type AG-TFET [7]

right half (10 nm) is controlled by the planar gate. The cross-sectional area of the GAA channel is 5×5 nm^2 and that of the planar channel is 15×20 nm^2. The effective oxide thickness is 1.3 nm (to meet the International Technology Roadmap for Semiconductors: ITRS), and the gate work function is 4.72 eV (to meet TiN). The doping concentrations of the p-type source, the n-type drain, and the low-doped n-type channel are 1×10^{20}, 1×10^{19}, and 1×10^{16} cm^{-3}, respectively.

Figure 7.22 compares the transfer characteristics of the n-channel AG-TFET to those of the gate-all-around (GAA) and the planar TFET. The simulated on-state current (I_{on}) in a GAA TFET is $\mathbf{2.11 \times 10^{-5}}$ **A/μm** at $V_g = 2$ V, and the off-state current (I_{off}) in a planar TFET is 1.51×10^{-15} A/μm at $V_g = 0$ V. The SS of the GAA TFET and the planar TFET is 61 mV/dec and 124 mV/dec, respectively. Therefore, the AG-TFET combines the advantages of both structures, with a 2679-fold higher I_{on} than that of the planar TFET and a 476-fold lower I_{off} than that of the GAA TFET. When this asymmetric-gate architecture is used in the TFET, the minimum SS is $\mathbf{42}$ **mV/dec**, the average SS is 45 mV/dec (determined over three decades of I_d), and the maximum I_{on}/I_{off} ratio is 10^{10}.

Figures 7.23 and 7.24 present simulated energy band diagrams of the AG-TFET in the on-state ($V_g = 2$ V) and the off-state ($V_g = -0.5$ V), respectively. The energy band of the right half-channel of the AG-TFET will be effectively shifted down as the full GAA channel in the on-state. When the bias voltage is large enough to reduce the barrier width, electrons tunnel from the valence band of the source side to the conduction band of the channel side. In a TFET, the triangular barrier width is the screening tunneling length (λ). The screening lengths of a GAA and a planar structure, λ_1 and the λ_2, respectively, are given by

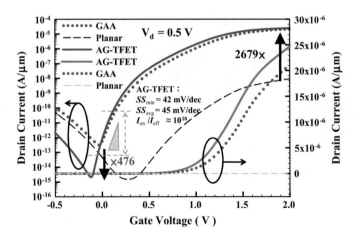

Fig. 7.22 Transfer characteristics of the n-type AG-TFET [7]

Fig. 7.23 Simulated energy band diagram of n-type AG-TFET in channel direction in on-state ($V_g = 2$ V, $V_d = 0.5$ V) [7]

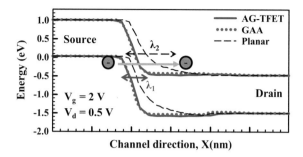

Fig. 7.24 Simulated energy band diagram of n-type AG-TFET in channel direction in off-state ($V_g = -0.5$ V, $V_d = 0.5$ V) [7]

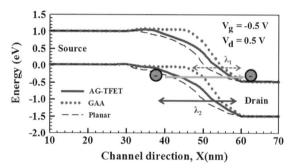

$$\lambda_1 = \sqrt{\frac{\varepsilon_{Si}}{4\varepsilon_{ox}} T_{Si} T_{ox}} \tag{7.6}$$

$$\lambda_2 = \sqrt{\frac{\varepsilon_{Si}}{\varepsilon_{ox}} T_{Si} T_{ox}} \tag{7.7}$$

where ε_{si} and ε_{ox} are the dielectric constants of silicon and oxide, respectively. T_{si} and T_{ox} are the thickness of the silicon and the oxide, respectively. The value of λ_1 is half that of λ_2. Accordingly, the tunneling probability in the GAA structure is higher than in the planar structure. Figure 7.23 reveals that using the GAA structure on the source side increases the I_{on} of a TFET. When the TFET is operated in the off-state, electrons tunnel from the valance band of the channel side to the conduction band of the drain side, producing a leakage current in the TFET. Thus, Fig. 7.24 indicates that the planar structure that is used at the drain side reduces the I_{off} of the TFET owing to its large screening tunneling length.

Figures 7.25 and 7.26 present the BTBT generation rates in the channel direction at the center of the channel in the AG-TFET at $V_g = 2$ V and $V_g = -0.5$ V, respectively. The source tunneling junction is formed at X = 30 nm (NW region), and the drain tunneling junction is generated at X = 50 nm (planar region). As expected, the BTBT generation rate is highest on the shortest tunneling path, as presented in Figs. 7.23 and 7.24. Therefore, the generation rate in the planar structure is six orders of magnitude smaller than that in the GAA structure.

Fig. 7.25 Rate of
band-to-band tunneling
generation in channel
direction in n-type AG-TFET
at $V_g = 2$ V, $V_d = 0.5$ V [7]

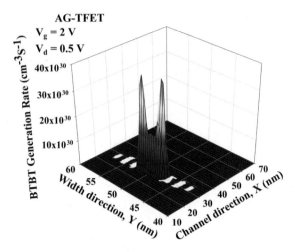

Fig. 7.26 Rate of
band-to-band tunneling
generation in channel
direction in n-type AG-TFET
at $V_g = -0.5$ V, $V_d = 0.5$ V
[7]

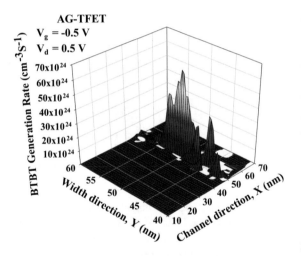

Figure 7.25 shows BTBT generation rate peaks close to the gate dielectric, as predicted by the screening length formulas 7.6 and 7.7. The peak in Fig. 7.26 is at the center of the channel, because the electric field is concentrated there.

The channel series resistance of the AG-TFET is lower than that of the GAA TFET, so the AG-TFET has a higher on-current than does the GAA TFET. Figure 7.27 compares the output characteristics of the AG-TFET, the GAA TFET, and the planar TFET. Clearly, the AG-TFET has a higher saturation drain current than the GAA TFET because the AG-TFET has a lower series resistance. The AG-TFET has a lower off-current and *SS*, because its planar part has a larger screening length. The planar TFET has the largest screening length, and therefore the lowest drain current. Figure 7.28 plots the output characteristic of the AG-TFET

Fig. 7.27 Output
characteristics of n-type
AG-TFET, GAA TFET, and
planar TFET at $V_g = 2$ V [7]

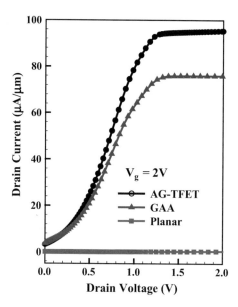

Fig. 7.28 Output
characteristics of n-type
AG-TFET at various gate
voltages. Effective channel
width is 20 nm [7]

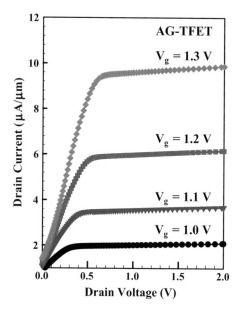

as a function of the gate voltage. Its low parasitic resistance and excellent drain
current saturation behavior reveal its potential for use in future low-power integrated circuits.

7.6   Summary of This Chapter

This chapter demonstrated the standard example of TFET simulation. The electrical properties indicate $SS < 60$ mV/dec., the ultra-low leakage current, I_{off} (in fA), and extremely high I_{on}/I_{off} ratio. The aforementioned simulation results show that TFET is very suitable for future application of ultra-low-power semiconductor device.

References

1. A.M. Ionescu, H. Riel, Tunnel field-effect transistors as energy-efficient electronic switches. Nature **479**, 329 (2011)
2. K. Jeon, W.Y. Loh, P. Patel, C.Y. Kang, J. Oh, A. Bowonder, C. Park, C.S. Park, C. Smith, P. Majhi, H.H. Tseng, R. Jammy, T.J. King Liu, C. Hu, Si tunnel transistors with a novel silicided source and 46 mV/dec swing. VLSI Tech. Symp. **121** (2010)
3. S. Richter, C. Sandow, A. Nichau, S. Trellenkamp, M. Schmidt, R. Luptak, K.K. Bourdelle, Q.T. Zhao, S. Mantl, Ω-gated silicon and strained silicon nanowire array tunneling FETs. IEEE Electr. Dev. Lett. **33**, 1535 (2012)
4. K. Boucart, A.M. Ionescu, Double-gate tunnel FET with high-κ gate dielectric. IEEE Trans. Electron Dev. **54**, 1725 (2007)
5. K.T. Lam, D. Seah, S.K. Chin, S.B. Kumar, G. Samudra, Y.C. Yeo, G. Liang, A simulation study of graphene-nanoribbon tunneling FET with heterojunction channel. IEEE Electr. Dev. Lett. **31**, 555 (2010)
6. Q. Huang, Z. Zhan, R. Huang, X. Mao, L. Zhang, Y. Qiu, Y. Wang, Self-depleted T-gate Schottky barrier tunneling FET with low average subthreshold slope and high I_{on}/I_{off} by gate configuration and barrier modulation. Tech. Digest of IEDM 13.2.1 (2011)
7. Y.R. Jhan, Y.C. Wu, M.F. Hung, Performance enhancement of nanowire tunnel field-effect transistor with asymmetry-gate based on different screening length. IEEE Electr. Dev. Lett. **34**, 1482 (2013)

Chapter 8
Extremely Scaled Si and Ge to L_g = 3-nm FinFETs and L_g = 1-nm Ultra-Thin Body Junctionless FET Simulation

8.1 Foreword

Huge efforts are put into CMOS scaling to push the limits of Moore's law. Semiconductor ICs manufacturing companies are currently ramping up 16-nm/14-nm FinFET processes, with 7 and 5 nm technology nodes just around the corner. As we approach sub-10-nm node technologies, different device models have been proposed and intensively researched to overcome the several critical challenges that arise due to the relentless scaling to ever small dimensions. Various approaches have been proposed and comprehensively explored to attenuate the impact of short-channel effect (SCE) on threshold voltage, drain-induced barrier lowering (DIBL), and subthreshold swing (SS). Leakage current (I_{off}) and electrostatics (gate control) become important factors of concern. High-k dielectrics and high mobility materials and various device architectures are explored extensively. There are a few papers that successfully address challenges in ultra-scaled gate length real devices [1, 2].

Finding a suitable semiconductor material for the sub-10-nm technology is the major challenge for the semiconductor researchers around the world. The materials that are investigated need to be compatible to the current CMOS technology adapted by the industry. The next-generation materials that possess similar qualities as that of Silicon which is cost efficient as well as that suits the industry requirement is very difficult to identify. The reliability of the newly investigated materials meets many challenges. The silicon and germanium technologies are more mature and are most researched for many decades now. Suitable continuous scaling of transistors hinders the development of high-quality junctions especially in sub-10-nm nodes where modifying the doping concentration becomes an even more strenuous process.

© Springer Nature Singapore Pte Ltd. 2018
Y.-C. Wu and Y.-R. Jhan, *3D TCAD Simulation for CMOS Nanoeletronic Devices*,
DOI 10.1007/978-981-10-3066-6_8

8.1.1 Challenges of Sub-10-nm Technology Node

Several challenges need to be addressed in sub-10-nm technology, and some of the important challenges are as follows: leakage current (I_{off}), I_{on}/I_{off} current ratio that determines the switching speed of the device and the threshold voltage roll off and variation, the effective mass of high mobility materials, quantum confinement, Fin width, and Fin height optimization. The short-channel effects such as SS DIBL and other effects. To put them all in a broader topic challenges that will arise can be classified into following divisions: physical challenges, material challenges, power and thermal challenges, technological challenges, and economical challenges. Also in sub-10-nm technology node, many other quantum effects will be prominent such as ballistic conduction, tunneling, and uncertainty principle.

8.1.2 Material Selection for Sub-10-nm Technology Node

(a) Silicon

Silicon is one of the most researched materials worldwide. It is abundance in nature, easy to handle, robustness, and cheaper cost made it a favorite candidature in semiconductor industry. For the past two decades, scientists around the world successfully devised many new ways to scale down Si-based devices. Hence, the Si technology is extremely mature and more reliable than any other semiconductor material.

(b) Germanium

The first transistor emerged from germanium almost seven decades ago. But it represents very small market today because of its instability with a lower bandgap energy compared to Si. On the other hand, Ge has higher electron and hole mobility. Thus, Ge devices can function at higher frequencies than Si devices. This makes Ge a promising candidate for sub-10-nm node. Also, Ge device technology is similar to that of current industrial Si technologies.

(c) III-V and 2D high mobility semiconductor materials

Other important materials are mainly III-V compound semiconductors composed of elements of group III (basically Al, Ga, and In) and elements of group V (basically N, P, As, and Sb). Among a total of 12 combinations, the combinations most likely to replace silicon include GaAs, InP, GaP, GaN, and InAs. Recently, there have been many new 2D materials being studied such as graphene and MoS_2. However, these higher mobility materials still face numerous problems including mass automotive production challenge, threshold voltage (V_{th}) control of nFET and pFET challenge, high off-sate leakage current (I_{off}), reliability challenge, and mass production cost challenge.

This main purpose of this chapter is to provide a logical understanding of the physical and electrical properties through the simulated results as the device dimension approaches gate length (L_g) of 3 nm.

8.2 Design Guideline of Sub-20-nm to 9-nm Gate Length Si FinFET of Wine-Bottle Channel

In this section, we propose the design guideline of sub-20-nm to 9-nm gate length (L_g) Si FinFET wine-bottle shape Fin structure. The real sub-20-nm experimental results show that taller and thinner FinFET has lower I_{off} and lower drain-induced barrier lowering (DIBL). Through TCAD simulation, we predict that the wine-bottle shape FinFET is promising future sub-10-nm FinFET with excellent gate control to eliminate short-channel effect (SCE) and sufficient volume for epitaxing low resistance raised source and drain (S/D) materials. The V_{th} and SS are all reasonable and insensitive ($\Delta V_{th} < 14$ mV, $\Delta SS < 3$ mV/dec) to various sizes of wine-bottle shape Fin structure. The I_{on} increases with Fin height (F_h), top Fin weight (F_{wt}) increases monotonically, and I_{off} is vice verse. The wine-bottle FinFET is promising for Moore's law that can extend to sub-10-nm L_g CMOS IC technology.

8.2.1 Device Structure and Sub-20-nm FinFET Experimental Data

Figure 8.1 shows simulation structure and important design parameters of wine-bottle FinFET. The electron and hole density of $L_g = 28$-nm Si FinFET are shown in Fig. 8.2.

8.2.2 Simulation Results and Discussion

Figure 8.3 shows $L_g = 28$ nm, $L_g = 20$ nm, $L_g = 12$ nm, and $L_g = 9$ nm I_d–V_g curves, respectively. The Sentaurus TCAD simulation tool was applied to perform 3D simulations, which included the coupled drift-diffusion (DD) and density-gradient (DG) solving model with quantum effects. The bandgap narrowing, band-to-band tunneling, and SRH recombination models along with doping-dependent models are also considered. The mobility model used in device simulation is according to Matthiessen's rule including surface acoustic phonon scattering, surface roughness scattering, and bulk mobility with doping-dependent modification effect, respectively. For $L_g = 12$-nm and $L_g = 9$-nm FinFET, the F_h can be reduced around 20–30 nm for mechanical strength. Nevertheless, the ion must use strained Si and raised S/D engineering for high I_{on}.

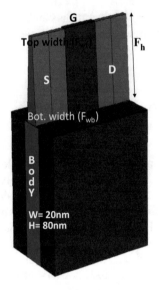

L_g(nm)	28nm	20nm	12nm	9nm
EOT(nm)	0.5	0.5	0.3	0.3
F_h(nm)	46-60	46-60	50	50
F_{wt}(nm)	2-6	2-6	3	3
F_{wb}(nm)	12	12	5	5
S/D Dop.	8E19	8E19	8E19	8E19
Ch. Dop.	8E18	8E18	5E18	5E18

Fig. 8.1 Device structure and parameters of simulated wine-bottle FinFET, with Fin height (F_h), top Fin width (F_{wt}), and bottom Fin width (F_{wb})

Figure 8.4 shows simulated linear I_{ds}–V_{gs} sub-20-nm FinFET plots versus (a) F_h and (b) F_{wt}. The I_{ds} is highly depending on F_{wt} rather than F_h. The ion increases with the F_{wt} increasing.

Figure 8.5 shows simulated linear I_{ds}–V_{gs} curve of wine-bottle FinFET keep the trapezoidal Fin of (a) pFET and (b) nFET. The results reveal that the I_{on} can be increased (+18%) by using tall F_h and wide top Fin width (F_{wt}).

Figure 8.6 shows simulated 3D contour plot influence of F_h and F_{wt} for sub-20-nm wine-bottle nFinET and pFinET with V_{th}, I_{on}, and I_{off}, respectively. The V_{th} and SS (not shown) are all reasonable and insensitive values (V_{th} < 14 mV, SS < 3 mV/dec). The ion increases with Fin height (F_h), top Fin weight (F_{wt}) increases monotonically, and I_{off} is in opposite trend. Once achieving the target V_{th} and I_{off} values, the I_{on} can be increased by using larger F_{wt}.

Figure 8.7 shows simulated 3D contour plot influence of F_h and F_w for L_g = 20-nm wine-bottle nFinET and pFinET with V_{th}, I_{on}, and I_{off}, respectively. The V_{th} and SS are all reasonable and insensitive values (ΔV_{th} < 8 mV, ΔSS < 4 mV/dec). The V_{th} insensitivity reveals that the V_{tn} and V_{tp} can entirely adjust by using proper metal gate materials with different work functions. The I_{on} increases with Fin height (F_h), top Fin weight (F_{wt}) increases monotonically, and I_{off} is in opposite trend.

Figure 8.8a plots the simulated timing characteristics of a Si CMOS inverter circuit of simulated L_g = 12-nm wine-bottle FinFETs. The T_{hl} is 0.89 ps, and T_{lh} is 1.8 ps. Figure 8.8b plots simulated SRAM characteristics with signal noise margin (SNM) of 160 mV.

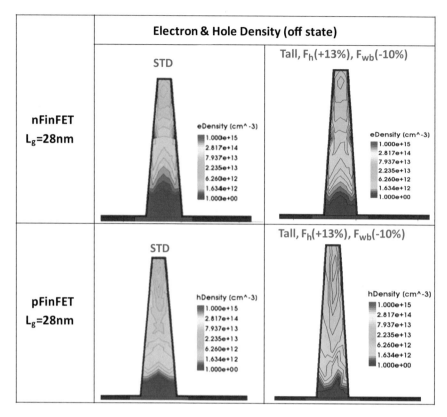

Fig. 8.2 Simulation results of electron and hole density at off-state of (**a**) sub-20 nm n-type and (**b**) p-type FinFET. The tall FinFET has lower electron and hole density distribution than STD FinFET

8.3 Study of Silicon L_g = 3-nm Bulk IM, AC, and JL FinFET

Methods are taken to reduce short-channel effects in traditional modes of operation such as inversion-mode (IM). In future, they may be replaced by other new modes of operation such as junctionless-mode (JL) of operation which is being researched widely nowadays. To obtain higher on-state current (I_{on}), JL transistors are heavily doped which leads to adverse effects on transport properties due to severe impurity scattering. Research study on standard IM and JL FinFETs shows that ultra-scaled FinFETs are inherently more sensitive to variability than standard devices and will pose significant challenges in post-CMOS technology. This chapter begins with the brief introduction to the challenges posed in the sub-10-nm technology. A general introduction about materials of current technology and prospective technology is analyzed by considering the industrial trend. The operation of device in sub-10-nm

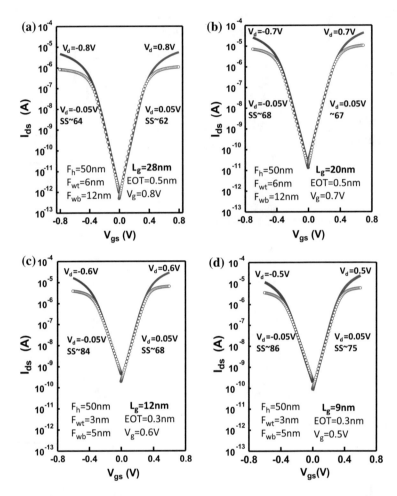

Fig. 8.3 Simulated I_{ds}–V_{gs} of (**a**) L_g = 28 nm, (**b**) L_g = 20 nm, (**c**) L_g = 12 nm, and (**d**) L_g = 9 nm wine-bottle FinFET

node will be completely different from that of the devices in higher technology nodes. The sub-10-nm technology devices will be more strictly adhering to the laws of quantum physics and important quantum confinement phenomenon, and size-dependent properties will come to effect more severely in sub-10-nm node. Hence, it is important to compare and analyze the performance of the conventional inversion-mode of operation along with the other modes of operation such as accumulation-mode and junctionless-mode. We examine the performance of the optimized 3-nm FinFET with homogeneous source and drain doping concentration in inversion-mode (IM), accumulation-mode (AC), and junctionless-mode (JL) operation. The transfer and output characteristics in IM, AC, and JL modes of simulated sub-5-nm technology node devices are discussed in detail. In addition,

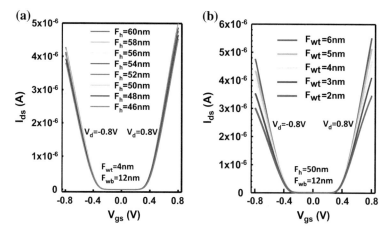

Fig. 8.4 Linear I_{ds}–V_{gs} sub-20-nm FinFET plots versus (**a**) F_h and (**b**) F_{wt}. The I_{ds} is highly increasing with the F_{wt}

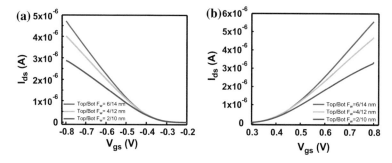

Fig. 8.5 Linear I_{ds}–V_{gs} sub-20-nm FinFET trapezoidal Fin of (**a**) pFET and (**b**) nFET

for each case, we interpret the 3D electron density mesh plots. The device performances such as the drain-induced barrier lowering, subthreshold slope, and on/off current ratio have also been estimated. This chapter serves as only a design guideline and in future with more ab initio and first principle-based models can be incorporated in the device physics for more accurate results.

In this section, we investigated the device performance of the optimized 3-nm gate length (L_g) bulk silicon FinFET device using 3D quantum transport device simulation. By keeping source and drain doping constant and by varying only the channel doping, the simulated device is made to operate in three different modes such as **inversion-mode (IM), accumulation-mode (AC), and junctionless-mode (JL)**. The excellent electrical characteristics of the 3-nm gate length Si-based bulk FinFET device were investigated. The subthreshold slope values (SS ∼ 65 mV/dec) and drain-induced barrier lowering (DIBL < 17 mV/V) are analyzed in all three IM, AC, and JL modes of bulk FinFET with $|V_{th}|$ ∼ 0.31 V.

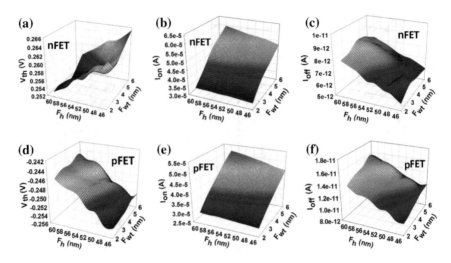

Fig. 8.6 (**a–c**) are 3D plots of L_g = 28-nm nFET with V_{th}, I_{on}, and I_{off}. (**d–f**) are 3D plots of pFET with V_{th}, I_{on}, and I_{off}, respectively. The F_{wb} is fixed at 12 nm

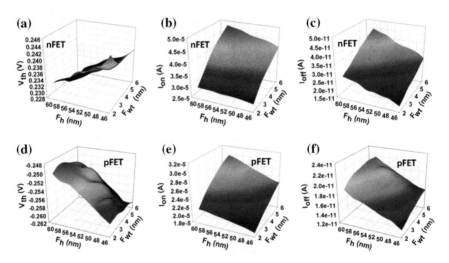

Fig. 8.7 (**a–c**) are 3D plots of L_g = 20-nm nFinET with V_{th}, I_{on}, and I_{off}, respectively. (**d–f**) are 3D plots of pFET with V_{th}, I_{on}, and I_{off}, respectively. F_{wb} is fixed at 12 nm

Furthermore, the threshold voltage (V_{th}) of the bulk FinFET can be easily tuned by varying the work function (WK). This research reveals that Moore's law can continue up to 3-nm nodes.

The simulated device structure and the table of important parameters used in the device simulation are given in Fig. 8.9. We applied equivalent oxide thickness (EOT) of 0.3 nm. The gate length (L_g) is 3 nm, and the Fin width (F_w) and the Fin

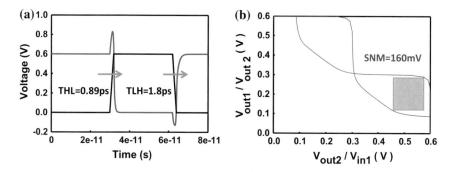

Fig. 8.8 Simulated L_g = 12-nm FinFET (**a**) inverter timing characteristics and (**b**) SRAM characteristics with signal noise margin (SNM) of 160 mV

height (F_h) are also the same ($F_w = F_h = 3$ nm). The doping concentrations of source/drain in all three modes (IM, AC, JL) of bulk FinFET devices are set to 1.0×10^{20} cm^{-3} for both n-type and p-type transistors. The channel concentration of JL bulk FinFET is set to 1.0×10^{20} cm^{-3}. The channel concentration of IM and AC bulk FinFET is set to 1.0×10^{18} cm^{-3}. Arsenic and boron are used as dopants in device simulation. The bulk doping concentration for the FinFET is 5×10^{18} cm^{-3}, which can be implemented easily by usual well doping implantation. A constant V_{th} value was maintained for both nFET ($V_{th} \sim 0.31$ V) and pFET ($V_{th} \sim -0.31$ V) in all three modes of operation. The work function used for n-type IM, AC, and JL modes is 4.40, 4.41, and 4.55 eV, respectively. Similarly, the work function for p-type IM, AC, and JL modes is 4.80, 4.81, and 4.69 eV, respectively. Precise numerical results of the simulated nanoscale device are obtained by solving 3D quantum transport equations provided by Synopsys Sentaurus version 2014. In quantum transport equations, a density-gradient model is used in the simulation. The bandgap narrowing model and Shockley–Read–Hall recombination with doping-dependent model are also considered. The mobility model used in device simulation is according to Matthiessen's rule.

$$R_{SRH} = \frac{np - n_{i,eff}^2}{\tau_p(n + n_1) + \tau_n(p + p_1)} \tag{8.1}$$

R_{srh} is the carrier composite item of Shockley–Read–Hall; τ_p and τ_n are lifetimes of electrons and holes; $n_{i,eff}$ is the effective intrinsic concentration; and n_1 and p_1 are constants of defect charge.

$$p = N_v F_{1/2}\left(\frac{E_{F,p} - E_v - \Lambda_p}{kT_p}\right) \tag{8.2}$$

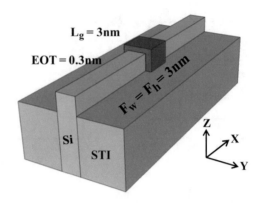

Device Mode	Junctionless	Accumulation	Inversion
Source & Drain Doping Concentration	N : 1x10²⁰ cm⁻³ P : 1x10²⁰ cm⁻³	N :1x10²⁰ cm⁻³ P :1x10²⁰ cm⁻³	N :1x10²⁰ cm⁻³ P :1x10²⁰ cm⁻³
Channel Doping Concentration		N : 1x10¹⁸ cm⁻³,N-Type P : 1x10¹⁸ cm⁻³,P-Type	N : 1x10¹⁸ cm⁻³,P-Type P : 1x10¹⁸ cm⁻³,N-Type
Substrate Doping Concentration	N : 5x10¹⁸ cm⁻³,P-Type P : 5x10¹⁸ cm⁻³,N-Type	N : 5x10¹⁸ cm⁻³,P-Type P : 5x10¹⁸ cm⁻³,N-Type	N : 5x10¹⁸ cm⁻³,P-Type P : 5x10¹⁸ cm⁻³,N-Type

Fig. 8.9 Device structure and important parameters of simulated 3-nm gate length (L_g) IM, AC, and JL Si bulk FinFET [3]

$$n = N_c F_{1/2} \left(\frac{E_{F,n} - E_c - \Lambda_n}{kT_n} \right) \tag{8.3}$$

p and n are concentrations of hole and electron, respectively; $F_{1/2}$ is Fermi–Dirac integral; N_c and N_v are effective densities of states of conduction band and valence band, respectively; and T_p and T_n are temperatures of hole and electron.

$$\Lambda_p = -\frac{\gamma \hbar^2}{12 m_p} \left[\nabla^2 \ln p + \frac{1}{2} (\nabla \ln p)^2 \right] \tag{8.4}$$

$$\Lambda_n = -\frac{\gamma \hbar^2}{12 m_n} \left[\nabla^2 \ln n + \frac{1}{2} (\nabla \ln n)^2 \right] \tag{8.5}$$

m_p and m_n are effective mass of hole and electron, respectively, and E_{trap} is the difference between defect energy level and intrinsic energy level. In addition, the mobility model in the device simulation is in accordance with the following Matthiessen's rule:

$$\frac{1}{\mu} = \frac{D}{\mu_{surf\_aps}} + \frac{D}{\mu_{surf\_rs}} + \frac{1}{\mu_{bulk\_dop}} \tag{8.6}$$

In $D = \exp(x/l_{crit})$, x is the distance from the interface, and l_{crit} is the fitting parameter. The mobility is composed of three kinds of phenomena, such as acoustic phonon scattering (μ_{surf_aps}), surface roughness scattering (μ_{surf_rs}), and bulk mobility with doping-dependent modification (μ_{bulk_dop}).

Results and Discussion

Left-hand side plots of Figs. 8.10, 8.11, and 8.12 show the I_d–V_g curves of the n-type and p-type devices of interest, in which the linear threshold voltage (V_{th}) is all adjusted to approximately ±300 mV for fair comparison. In the proposed n-type IM, AC, and JL bulk FinFET, the saturation current (at V_g = 0.7 V, V_d = 1 V) is 2.52×10^{-4} A/μm, 2.54×10^{-4} A/μm, and 2.32×10^{-4} A/μm, respectively. For p-type IM, AC, and JL bulk FinFET, the saturation current is 2.24×10^{-4}, 2.25×10^{-4}, and 2.26×10^{-4} A/μm, respectively. The SS for n-type IM, AC, and JL modes is, respectively, 78.74, 78.79, and 77.37 mV/dec. The SS for p-type IM, AC, and JL modes is 67.63, 67.65, and 62.28 mV/dec, respectively. The DIBL, defined as the difference in V_{th} between V_d = 0.05 V and V_d = 0.7 V, for n-type IM, AC, and JL modes, equals only 16.04, 16.17, and 26.80 mV/V, respectively.

The similar performances are also achieved in p-type IM, AC, and JL bulk FinFET (29.80, 31.89 mV/V, and 40.20 mV/V). The DIBL and SS numerical values are tabulated in Table 8.1. As the F_w and F_h are reduced to 3 nm, the SS and DIBL approach to their ideal value (60 mV/dec and 0 mV/V) in the simulated results.

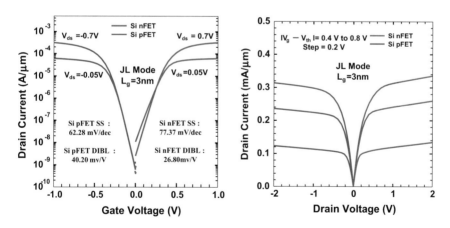

Fig. 8.10 I_d–V_g of 3-nm gate length (L_g) for n-type and p-type Si bulk FinFET operating in JL mode with SS and DIBL values shown *inset* and I_d–V_d of 3-nm gate length (L_g) for n-type and p-type Si bulk FinFET operating in JL mode, with overdrive voltage $|V_{ov}| = |V_g - V_{th}|$ [3]

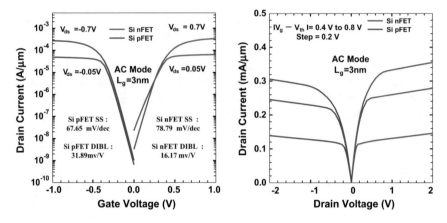

Fig. 8.11 I_d–V_g of 3-nm gate length (L_g) for n-type and p-type Si bulk FinFET operating in AC mode with SS and DIBL values shown *inset* and I_d–V_d of 3-nm gate length (L_g) for n-type and p-type Si bulk FinFET operating in AC mode, with overdrive voltage $|V_{ov}| = |V_g - V_{th}|$ [3]

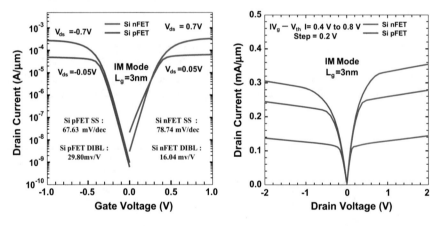

Fig. 8.12 I_d–V_g of 3-nm gate length (L_g) for n-type and p-type Si bulk FinFET operating in IM with SS and DIBL values shown *inset* and I_d–V_d of 3-nm gate length (L_g) for n-type and p-type Si bulk FinFET operating in IM, with overdrive voltage $|V_{ov}| = |V_g - V_{th}|$ [3]

It is noteworthy that the off-state current is all low in IM, AC, and JL modes of Si bulk FinFET owing to extensively scaled nanofin.

Right-hand side plots of Figs. 8.10, 8.11, and 8.12 show the output characteristic curves of Si FinFET. It is very clear that simulated IM and AC devices have almost similar I_d–V_d output characteristic curves.

Figure 8.13a, b compares on-state (V_{gs} = 1 V) and off-state (V_{gs} = 1 mV) electron density distribution at the 3D cross sections of the 3-nm nanofin n-type Si bulk JL-FinFET. The conduction path is located at the middle of the nanofin as expected. Figure 8.13c–f compares on-state (V_{gs} = 1 V) and off-state (V_{gs} = 1 mV)

Table 8.1 Important numerical values of simulated 3-nm gate length IM, AC, and JL Si bulk FinFETs [3]

Device mode	Junctionless	Accumulation	Inversion
Work function (eV)	N: 4.55 P: 4.60	N: 4.40 P: 4.80	N: 4.40 P: 4.80
V_{th} (~0.31 V)	N: 0.3057 P: −0.2969	N: 0.2987 P: −0.2987	N: 0.3000 P: −0.3019
SS (mV/dec)	N: 77.37 P: 62.28	N: 78.79 P: 67.65	N: 78.74 P: 67.63
DIBL (mv/V)	N: 26.80 P: 40.20	N: 16.17 P: 31.89	N: 16.04 P: 29.80
I_{on} (A/μm)	N: 2.32×10^{-4} P: 2.26×10^{-4}	N: 2.54×10^{-4} P: 2.24×10^{-4}	N: 2.522×10^{-4} P: 2.240×10^{-4}

electron density distribution at the 3D cross sections of the 3-nm nanofin n-type Si bulk IM-FinFET and AC-FinFET, respectively. Notably, the on-state and off-state results of 3-D eDensity distribution from quantum transport simulation demonstrate that the device can be scaled down to a physical limit of 3-nm node. The current conduction in all three (JL, AC, and IM) modes is almost similar because the carriers fully occupy 3-nm nanofin cross section. The electrons are more concentrated at the middle topside of channel in IM, AC, and JL bulk FinFET as better controllability by gate is achieved with $L_g = F_w = F_h = 3$ nm.

In summary, we have performed various analyses in the 3-nm gate length bulk silicon FinFET operating in inversion-mode (IM), accumulation-mode (AC), and junctionless-mode (JL). The observed transfer characteristics, output characteristics, and electron density distribution results of the 3D quantum transport device simulation reveal the fact that all the three IM, AC, and JL modes of operation are perfectly feasible even at 3-nm gate length. Thus, it enables the bulk FinFET devices to be scaled down to its least possible physical limits obeying Moore's scaling law.

8.4   Study of Germanium L_g = 3-nm Bulk FinFET

In this section, the Synopsys Sentaurus TCAD 2014 version 3D device simulation is used to show the performances of n-type and p-type 3-nm bulk Ge FinFET of IM-FET, AC-FET, and JL—FET. The simulated bulk Ge FinFET device exhibits better short-channel characteristics, including drain-induced barrier lowering (DIBL < 10 mV/V) and subthreshold slope (SS ~ 64 mV/dec). Electron density distributions in on-state and off-state also show that the simulated devices have better I_{on}/I_{off} ratios.

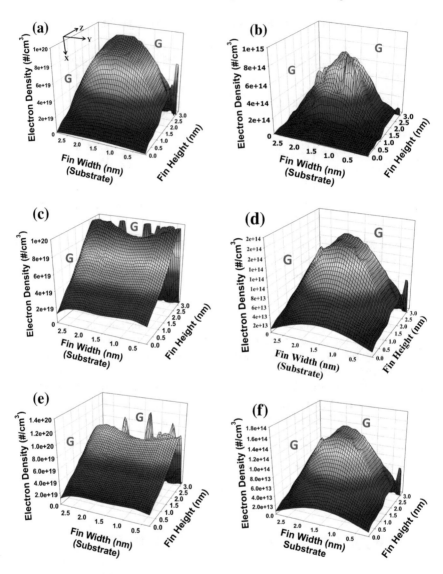

Fig. 8.13 3D mesh plot for electron density distributions in the 3-nm gate length (L_g) n-type Si bulk FinFET in (**a**) JL on-state (V_{gs} = 1 V) and (**b**) JL off-state (V_{gs} = 1 mV); (**c**) IM on-state and (**d**) IM off-state; and (**e**) AC on-state and (**f**) AC off-state [3]

Simulation Method

The simulated device structure and the table of important parameters used in the device simulation are given in Fig. 8.14. We applied equivalent oxide thickness (EOT) of 0.3 nm. The gate length (L_g) is 3 nm, and the Fin width (F_w) and the Fin height (F_h) are also the same ($F_w = F_h$ = 3 nm). The doping concentrations of source/drain in all three modes (IM, AC, JL) of bulk FinFET devices are set to 1.0×10^{20} cm^{-3} for both n-type and p-type transistors. The channel concentration of bulk JL-FinFET is set to 1.0×10^{20} cm^{-3}. The channel concentration of IM and AC bulk FinFET is set to 1.0×10^{18} cm^{-3}. Arsenic and boron are used as dopants in device simulation. The bulk doping concentration for the FinFET is 5×10^{18} cm^{-3}. A constant V_{th} value was maintained for both nFET ($V_{th} \sim 0.31$ V) and pFET ($V_{th} \sim -0.31$ V) in all three modes of operation. The work function used for n-type IM, AC, and JL modes is 4.40, 4.41, and 4.40 eV, respectively. Similarly, the work function used for p-type IM, AC, and JL modes is 4.37, 4.37, and 4.33 eV, respectively.

Results and Discussion

I_d–V_g curves of Si and Ge 3-nm bulk FinFETs are as shown in Figs. 8.15a, 8.16a, and 8.17a. The I_{sat} of Ge IM, AC, and JL nFET is around 3.0×10^{-4} A/µm. The I_{sat} of Ge IM, AC, and JL pFET is around 3.3×10^{-4} A/µm. The I_{sat} of Si IM, AC, and JL nFET is around 3.1×10^{-4} A/µm. The I_{sat} of Si IM, AC, and JL pFET is around 2.8×10^{-4} A/µm.

The SS values for Ge IM nFET and Si IM nFET are 64.38 and 78.74 mV/dec. The DIBL value for Ge IM pFET and Si IM pFET is 5.43 and 29.80 mV/V, which is ~5× times higher for Si compared to Ge. The simulated AC mode of Ge nFET

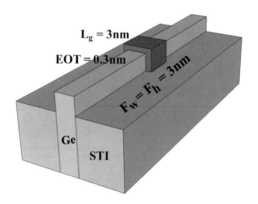

Device Mode	Junctionless	Accumulation	Inversion
Source & Drain Doping Concentration	N : 1x10²⁰ cm⁻³ P : 1x10²⁰ cm⁻³	N :1x10²⁰ cm⁻³ P :1x10²⁰ cm⁻³	N :1x10²⁰ cm⁻³ P :1x10²⁰ cm⁻³
Channel Doping Concentration		N : 1x10¹⁸ cm⁻³,N-Type P : 1x10¹⁸ cm⁻³,P-Type	N : 1x10¹⁸ cm⁻³,P-Type P : 1x10¹⁸ cm⁻³,N-Type
Substrate Doping Concentration	N : 5x10¹⁸ cm⁻³,P-Type P : 5x10¹⁸ cm⁻³,N-Type	N : 5x10¹⁸ cm⁻³,P-Type P : 5x10¹⁸ cm⁻³,N-Type	N : 5x10¹⁸ cm⁻³,P-Type P : 5x10¹⁸ cm⁻³,N-Type

Fig. 8.14 Device structure and important parameters of simulated 3-nm gate length (L_g) IM, AC, and JL germanium bulk FinFET

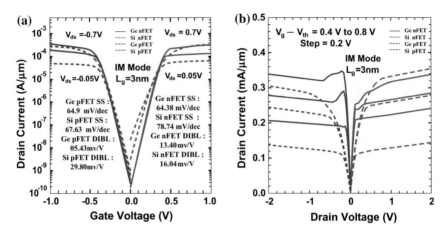

Fig. 8.15 (a) I_d–V_g of 3-nm gate length (L_g) for n-type and p-type Ge bulk FinFET operating in IM with SS and DIBL values shown *inset* and (b) I_d–V_d of 3-nm gate length (L_g) for n-type and p-type Ge bulk FinFET operating in IM, with overdrive voltage $|V_{ov}| = |V_g - V_{th}|$

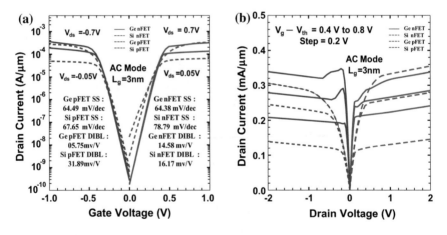

Fig. 8.16 (a) I_d–V_g of 3-nm gate length (L_g) for n-type and p-type Ge bulk FinFET operating in AC mode with SS and DIBL values shown *inset* and (b) I_d–V_d of 3-nm gate length (L_g) for n-type and p-type Ge bulk FinFET operating in AC mode, with overdrive voltage $|V_{ov}| = |V_g - V_{th}|$

and Si nFET obtains a SS value of 64.38 and 78.79 mV/dec, respectively, and DIBL values of Ge AC pFET and Si AC pFET are 5.75 and 31.89 mV/V, respectively.

I_d–V_g of Ge JL nFET achieves almost ideal SS value of 65.92 mV/dec, and Si JL nFET has SS value of 77.37 mV/dec. The DIBL value of Ge JL pFET and Si JL pFET is 8.38 and 40.20 mV/V, respectively. It is noteworthy that the off-state

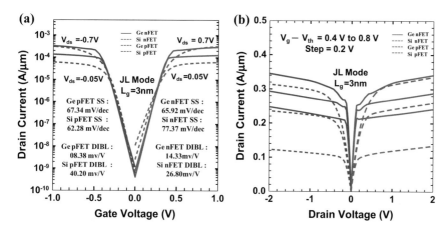

Fig. 8.17 (**a**) I_d–V_g of 3-nm gate length (L_g) for n-type and p-type Ge bulk FinFET operating in JL mode with SS and DIBL values shown *inset* and (**b**) I_d–V_d of 3-nm gate length (L_g) for n-type and p-type Ge bulk FinFET operating in JL mode, with overdrive voltage $|V_{ov}| = |V_g - V_{th}|$

current is all low in IM, AC, and JL modes of both bulk FinFETs owing to extensively scaled nanofin.

Figs. 8.15b and 8.16b show the I_d–V_d curves of IM and AC Ge pFET with an anomalous kink-effect behavior. It could be explained by the decrease in hole quantum capacitance (C_q) in IM and AC modes which in turn degrades gate to channel capacitance in 3-nm nanofin owing to the light hole effective mass ($m_p^* \sim 0.04m_0$) of Ge.

On the other hand, Fig. 8.17b shows that JL Ge pFET has almost negligible kink effect. It is significant to note that the IM, AC, and JL modes of Si pFET show no such irregularity, in the I_d–V_d characteristics. It must be noted that quantum confinement effect (QCE) is involved in this case and the principle is yet to be clarified on further research study.

The comparisons of electron density distributions and electrostatic potentials of 3D cross sections of 3-nm nanofin n-type Ge IM and JL bulk FinFETs are as shown in Fig. 8.18, where the conduction path is located at the middle of the nanofin. The electron density distributions of n-type Ge bulk FinFET in on-state (V_{gs} = 1 V) and off-state (V_{gs} = 1 mV) are as shown in Fig. 8.18a, b.

Table 8.2 summarizes the key parameters used in the quantum transport simulation and significant results which demonstrate high-performance Ge bulk FinFET at ultra-scaled L_g of 3-nm.

In summary, the L_g = 3-nm Ge bulk FinFETs in operations under IM mode, AC mode, and JL mode are analyzed in this section. Ge FinFET has better electrical performance compared to Si FinFET. However, smaller effective mass of Ge results in degradation of quantum capacitance compared to a standard 3-nm gate length Si

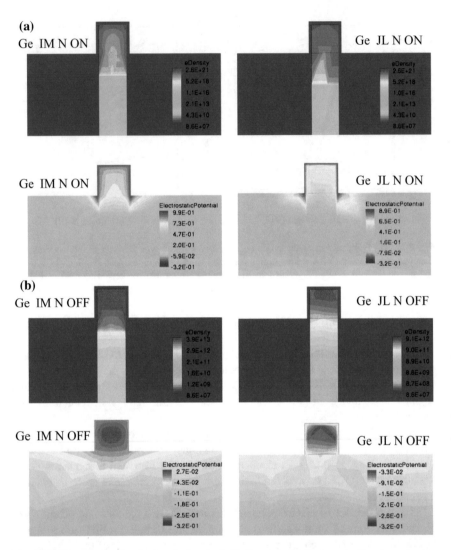

Fig. 8.18 Electron density (*top*) and electric field (*bottom*) distributions in the channel when n-type JL and IM devices operate at (**a**) on-state (V_{gs} = 0.7 V) and (**b**) off-state (V_{gs} = 1 mV) with $L_g = F_w = F_h$ = 3 nm and EOT = 0.3 nm

device. The electron density distribution reveals the fact that with an optimized 3-nm nanofin, charge carriers fully occupy the Fin region in all three modes (JL, AC and IM) of operation. The observed transfer, output characteristics iterates the fact that even at 3-nm L_g high-performance Ge bulk FinFET is feasible with all three IM, AC, and JL modes of operation for future sub-5-nm device applications.

Table 8.2 Important numerical values of simulated $L_g = 3$ nm

Device mode	Junctionless	Accumulation	Inversion
Work function (eV)	N: 4.40 P: 4.33	N: 4.41 P: 4.37	N: 4.40 P: 4.37
V_{th} (~ 0.31 V)	N: 0.2964 P: −0.3061	N: 0.3181 P: −0.2996	N: 0.3194 P: −0.3023
SS (mV/dec)	N: 65.92 P: 67.34	N: 64.38 P: 64.49	N: 64.38 P: 64.90
DIBL (mv/V)	N: 14.33 P: 08.38	N: 14.58 P: 05.75	N: 13.40 P: 05.43

8.5   Study of Silicon and Germanium UTB-JL—FET with Ultra-Short Gate Length = 1 and 3 nm

The next part is the introduction of simulation of ultra-thin body junctionless FET (UTB-JL—FET) of Si and Ge with $L_g = 1$ nm and $L_g = 3$ nm, which is coupled with the drift-diffusion (DD) and density-gradient (DG) models for finding solutions. The simulation results indicate that the UTB structure is well suited for Si and Ge. By using the UTB structure, the short-channel device does not have to be in compliance with the equation of $T_{ch} = L_g/3$. In addition, the Ge UTB-JL—FET 6T-SRAM has a reasonable static noise margin (SNM) of 149 mV. The circuit simulation result shows that UTB-JL—FET can be used for the CMOS technology node of sub-5 nm.

Junctionless field-effect transistor (JL—FET) structure can circumvent afore-mentioned issues because the channel region of JL—FET has high doping concentrations and the same dopant type as source/drain regions. Owing to the special doping profile, JL—FET has many advantages such as (1) lower thermal budget which can integrate with high-k/metal gate easier than conventional MOSFETs, (2) longer effective channel length than conventional MOSFETs, (3) the body current which can avoid surface scattering, and (4) avoidance of complicated source/drain engineering. Therefore, JL—FET is a potential candidate for ultra-short-channel transistor. But JL—FET has turnoff problem due to high doping concentrations in channel region. To solve this problem, JL—FET needs ultra-thin body (UTB) structure to reach fully depleted channel region in off-state.

The UTB structure can provide quantum confinement effect in channel region which will increase energy bandgap, and this large bandgap can suppress leakage current. Consequently, an empirical rule of $T_{ch} = L_g/3$ has been used for the definition of transistor dimension. As transistor features are scaled, the drive current (I_d) is declined. Therefore, a high mobility material is necessary for sub-10-nm technology node. Germanium (Ge) is a potential candidate owing to its high mobility. The electron mobility of Ge is two times higher than Si, and the hole mobility of Ge is four times higher than Si.

We investigate the electrical performance of Si UTB-JL—FET compared to Ge with $L_g = 1$ nm and $L_g = 3$ nm by 3D simulations. The transistors and circuit

performances are discussed in detail. The simulation results reveal that Si and Ge JL
—FET with UTB (1 nm) structure can be employed in sub-5-nm CMOS technology nodes. Furthermore, this UTB structure can be achieved in the future
technology nodes by focused ion beam (FIB) or reactive-ion etching (RIE). Using
atomic layer chemical vapor deposition system (ALD) and chemical-mechanical
polishing (CMP) processes can perform high-k/metal gate in this UTB-JL—FET for
future application.

Simulation Method

Figure 8.19 displays the architecture of the UTB-JL—FET and the parameters used
in the simulation. The Synopsys TCAD simulator was employed to perform 3D
simulations, which included the coupled drift-diffusion (DD), density-gradient
(DG) model, bandgap narrowing, and quantum effects. A Si and a Ge were used in
the simulated channel material. The channel width is 10 nm. Because of quantum
confinement effect, this ultra-short-channel (L_g = 1 nm) device has normally off
characteristics. As UTB is employed, ultra-short-channel device does not need to
follow an empirical rule of $T_{ch} = L_g/3$, which is often used as a guideline to suppress short-channel effect. We have shown conduction band energy (EC) diagrams
of Si UTB-JL—FET with L_g = 1 nm in both off-state and on-state. In off-state, the
UTB structure builds a high EC level at gated region because of quantum mechanism bandgap shift. This energy barrier can block electrons which pass through
channel by thermal injection and direct tunneling. In on-state, owing to the absence
of energy barrier, the electrons pass through channel by ballistic transport in
JL—FET.

Results and Discussion

Figure 8.20a, b shows conduction band energy (E_c) diagrams of Si UTB-JL—FET
with L_g = 1 nm in off-state and on-state, respectively. In off-state, the UTB structure builds a high E_c level at gated region because of quantum mechanism bandgap
shift. This energy barrier can block electrons which pass through channel by

Fig. 8.19 Device structure
and important parameters of
simulated UTB-JL—FET
with coupled drift-diffusion
(DD) and density-gradient
(DG) model [4]

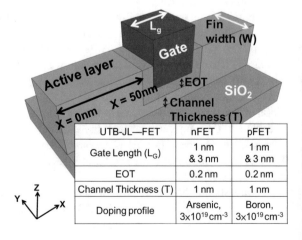

UTB-JL—FET	nFET	pFET
Gate Length (L_G)	1 nm & 3 nm	1 nm & 3 nm
EOT	0.2 nm	0.2 nm
Channel Thickness (T)	1 nm	1 nm
Doping profile	Arsenic, 3×10^{19} cm^{-3}	Boron, 3×10^{19} cm^{-3}

thermal injection and direct tunneling. In on-state, owing to the absence of energy barrier, the electrons pass through channel by ballistic transport in JL—FET.

Figure 8.21a, b shows the I_d–V_g characteristics of UTB-JL—FET with L_g = 1 nm in Si and Ge channel, respectively. Owing to the ultra-thin channel, this device has high I_{on}/I_{off} current ratio of 10^5 at V_g = 1 V. The SS is 100 mV/decade of Si pFET and 99 mV/decade of Si nFET, respectively. The DIBL is 225 mV/V of Si pFET and 222 mV/V of Si nFET, respectively. The SS is 96 mV/decade of Ge pFET and 93 mV/decade of Ge nFET, respectively. The DIBL is 196 mV/V of Ge pFET and 200 mV/V of Ge nFET, respectively. Even though this ultra-short-channel device does not follow an empirical rule of $T_{ch} = L_g/3$, the electrical properties can meet the industry requirements because of quantum confined UTB structure. The saturation current is 1.29×10^{-3} and 1.5×10^{-3} A/μm of Si nFET and Ge nFET, respectively. The saturation current is 0.82×10^{-3} A/μm and 1.08×10^{-3} A/μm of Si pFET and Ge pFET, respectively. The Ge nFET has a 16% higher saturation current (I_{sat}) than Si nFET, and the Ge pFET has a 32% higher I_{sat} than Si pFET.

Figure 8.22a, b shows the I_d–V_g characteristics of UTB-JL—FET with L_g = 3 nm in Si and Ge channels, respectively. UTB-JL—FET with L_g = 3 nm has lower SS and DIBL than L_g = 1 nm. The SS is 84 mV/decade of Si pFET and 83 mV/decade of Si nFET, respectively. It is worth noting that we have not used any strain engineering technique in our simulated device. The SS is 81 mV/decade of Ge pFET and 79 mV/decade of Ge nFET, respectively because Ge has higher channel mobility than Si channel. The Ge nFET has a 42% higher I_{sat} than Si nFET, and the Ge pFET has a 29% higher I_{sat} than Si pFET.

Figure 8.23a, b plots the timing characteristics of a CMOS inverter and static transfer characteristic curves of Si UTB-JL—FET with L_g = 1-nm 6T-SRAM cells, respectively. Figure 8.23c, d plots the timing characteristics of an CMOS inverter and

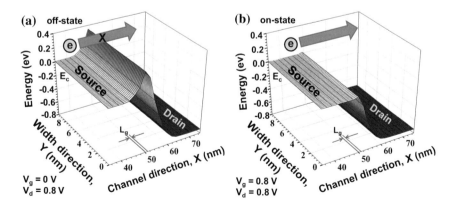

Fig. 8.20 Conduction band energy (E_c) diagrams of UTB-JL—FET with L_g = 1 nm and T = 1 nm in (**a**) off-state and (**b**) on-state. In off-state, a high E_c level at channel region and depletion width of 2.5 nm at both source and drain side to block leakage current [4]

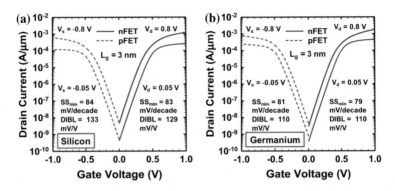

Fig. 8.21 I_d–V_g of UTB-JL—FET with channel thickness (T) is 1 nm, and gate length (L_g) is 1 nm for (**a**) silicon and (**b**) germanium channel [4]

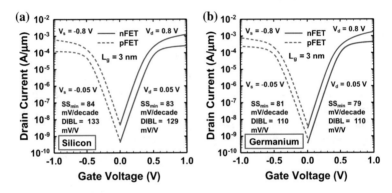

Fig. 8.22 I_d–V_g of UTB-JL—FET with channel thickness (T) is 1 nm, and gate length (L_g) is 3 nm for (**a**) silicon and (**b**) germanium channel [4]

static transfer characteristic curves of Ge UTB-JL—FET with L_g = 1-nm 6T-SRAM cells, respectively. Ge UTB-JL—FET shows lower delay time and larger static noise margin (SNM) than Si UTB-JL—FET. Ge UTB-JL—FET has large SNM value of 115 mV. These results demonstrate that ultra-short-channel device with UTB structure can be employed without following an empirical rule of T_{ch} = L_g/3.

Figure 8.24a, b plots the timing characteristics of a CMOS inverter and static transfer characteristic curves of Si UTB-JL—FET with L_g = 3-nm 6T-SRAM cells, respectively. Figure 8.24c, d plots the timing characteristics of an CMOS inverter and static transfer characteristic curves of Ge UTB-JL—FET with L_g = 3-nm 6T-SRAM cells, respectively. Ge UTB-JL—FET has a large SNM value of 149 mV.

In summary, Si UTB-JL—FET and Ge UTB-JL—FET with L_g = 1 nm and L_g = 3 nm were demonstrated successfully. The off-state leakage current can be reduced by quantum confinement effect. As UTB is employed, Si UTB-JL—FET and Ge UTB-JL—FET with L_g = 1 nm have high I_{on}/I_{off} current ratio of 10^5 at V_g = 1 V. Moreover, Ge UTB-JL—FET with L_g = 1 nm and L_g = 3 nm has

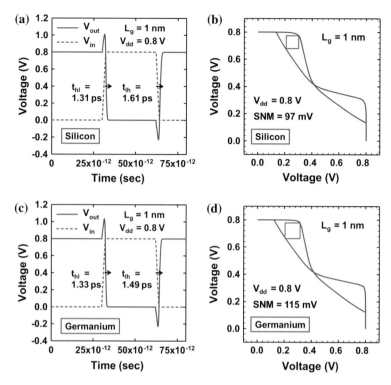

Fig. 8.23 **a** Timing characteristics of the input and output signals of a CMOS inverter for Si UTB-JL—FET with $L_g = 1$ nm. **b** Static transfer characteristic curves of Si UTB-JL—FET6T-SRAM cells. The definition of static noise margin (SNM) is the length of the side of the largest square that can be embedded inside the butterfly curve. **c** Timing characteristics of the input and output signals of an CMOS inverter for Ge UTB-JL—FET with $L_g = 1$ nm. **d** Static transfer characteristic curves of Ge UTB-JL—FET 6T-SRAM cells [4]

reasonable SNM that can meet the industry requirements. Using focus ion beam (FIB) or reactive-ion etching (RIE), this UTB recess channel structure can be achieved in sub-5-nm CMOS technology nodes. And this device can integrate high-k/metal gate by ALD and CMP. Finally, circuit performances reveal that UTB-JL—FET can be used in advanced logic ICs applications.

Summary of this chapter: Advanced examples of Si and Ge FinFET with $L_g = 3$ nm based on three different doping styles, such as inversion-mode (IM), accumulation-mode (AC), and junctionless-mode (JL) FinFETs, are proposed in the first half of this chapter, and the performances of all devices have been compared and analyzed. The simulation results reveal that the electric properties of inversion-mode, accumulation-mode, and junctionless-mode FinFETs are rather similar to each other in ultra-fine nanoscale channels. The UTB-JL—FET with L_g down to 1 nm has been proposed in the second half of this chapter. The simulation results reveal that extreme scaling UTB-JL—FET still maintains excellent electric properties. In addition, the characteristics of inverter and SRAM based on

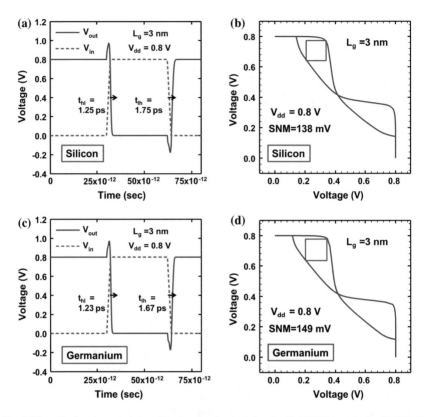

Fig. 8.24 **a** Timing characteristics of the input and output signals of a CMOS inverter for Si UTB-JL—FET with L_g = 3 nm. **b** Static transfer characteristic curves of Si UTB-JL—FET6T-SRAM cells. The definition of static noise margin (SNM) is the length of the side of the largest square that can be embedded inside the butterfly curve. **c** Timing characteristics of the input and output signals of a CMOS inverter for Ge UTB-JL—FET with L_g = 3 nm. **d** Static transfer characteristic curves of Ge UTB-JL—FET 6T-SRAM cells [4]

UTB-JL—FET with L_g = 1 nm are in compliance with the requirements of semi-conductor industry. Therefore, the results of this simulation can support the extension of Moore's law at least to 3-nm node.

References

1. S.D. Suk, M. Li, Y.Y. Yeoh, K.H. Yeo, J. K. Ha, H. Lim, H.W. Park, D.W. Kim, T.Y. Chung, K.S. Oh, W.S. Lee, Characteristics of sub 5 nm tri-gate nanowire MOSFETs with single and poly Si channels in SOI structure. VLSI Tech. Symp. 142 (2009)
2. S. Migita, Y. Morita, M. Masahara, H. Ota, Electrical performances of junctionless-FETs at the scaling limit (L_{ch} = 3 nm). Tech. Digest of IEDM, 8.6.1 (2012)

3. V. Thirunavukkarasu, Y.R. Jhan, Y.B. Liu, Y.C. Wu, Performance of inversion, accumulation, and junctionless mode n-type and p-type bulk silicon FinFETs with 3-nm gate length. IEEE Electr. Dev. Lett. **36**, 645 (2015)
4. Y.R. Jhan, V. Thirunavukkarasu, C.P. Wang, Y.C. Wu, Performance evaluation of silicon and germanium ultrathin body (1 nm) junctionless field-effect transistor with ultrashort gate length (1 nm and 3 nm). IEEE Electr. Dev. Lett. **36**, 654 (2015)

Appendix
Synopsys Sentaurus TCAD 2014 Version Software Installation and Environmental Settings

This is about the introduction of the latest Synopsys Sentaurus TCAD 2014 version (http://www.synopsys.com/tools/tcad/Pages/default.aspx) Copyright © 2015 Synopsys, Inc. The later version installation is similar. This simulation software (Sentaurus TCAD) can only be executed in Linux operating system, so this chapter will start with instructions on how to establish a Linux operating system environment under Windows environment following the sequence as shown below:

1. Downloading and installation of VMware Workstation.
2. Installation steps of VMware Workstation.
3. Installation of Synopsys Sentaurus TCAD software.

Although the installation illustration has some Chinese characters, we explain in English in all figure captions.

The first thing is about the recommendations for professional accessories of simulation PC host:

	CPU processor	Memory	Hard drive	Graphics card
Recommended accessories	Intel i7 and above	32G or above	2T or above	Independent graphics card
Minimum requirement of accessories	Intel i7	16G	1T	Independent graphics card

1. Downloading and installation of VMware Workstation

VMware Workstation (Copyright © 2015 VMware, Inc.) is a set of software which allows multiple operating systems to be executed on the same PC, while each operating system is equipped with an independent emulator just like an independent PC. Not only that, another version of the same operating system can be installed on the emulator without the need for additional hard drive partitioning; in addition, the virtual drive can be established in a portable hard drive or on a server, or even in a hard drive partition if necessary.

VMware Workstation can be used in conjunction with latest hardware to establish server in virtual machine and to establish desktop computer environmental

© Springer Nature Singapore Pte Ltd. 2018
Y.-C. Wu and Y.-R. Jhan, *3D TCAD Simulation for CMOS Nanoeletronic Devices*,
DOI 10.1007/978-981-10-3066-6

platform. Users will be allow to user multiple operating systems, including Linux, Windows or any other operating system on the same computer to execute application programs without the need for rebooting.

2. <u>Installation steps of VMware Workstation</u>

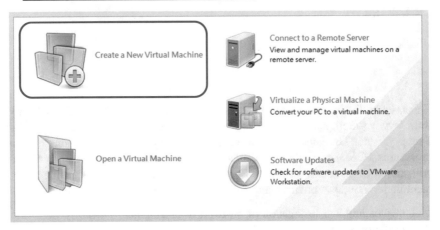

Step 1: Select [Create a New Virtual Machine]. (copyright © 2015 VMware, Inc).

Step 2: Select [Typical].

Step 3: Select the image of [CentOS-6.4]. This is because TCAD 2014 version can only be executed in Linux operating system.

Step 4: Select [Linux].

Step 5: Keep the original default value and directly go to the next step.

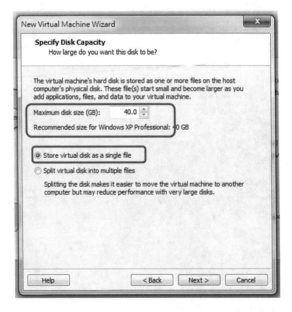

Step 6: Select the size of hard drive to be split for the Virtual Machine. **The recommended size is more than 100 GB.**

Step 7: Select [Customize Hardware] to enter the step of customized parameter setting.

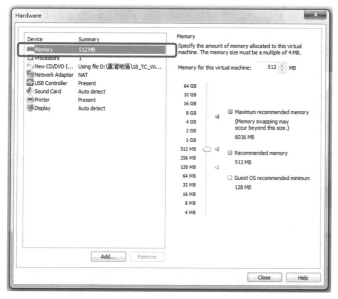

Step 8: The recommended memory is 16 GB or above.

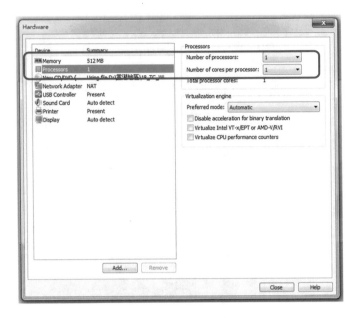

Step 9: Select [processor]. This step is mainly for determine (1) How many cores of CPU are to be assigned to this Virtual Machine; (2) How many threads per core are to be assigned to this Virtual Machine? The partition in this step should be in accordance with computer specification. A greater portion assigned to this Virtual Machine will accelerate the simulation.

Step 10: Press [Finish] after the setting is completed.

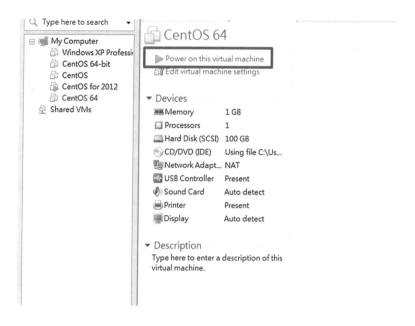

Step 11: Start [Virtual Machine] to begin with the installation of CentOS system.

Step 12: Press [Enter] upon entering this page. The user can install newest version in CentOS website.

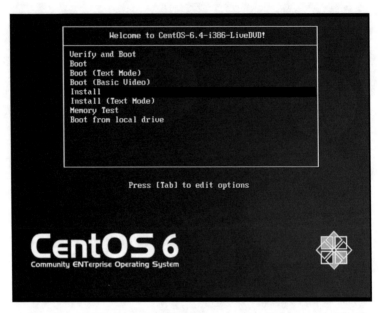

Step 13: Press [Install] to start the installation.

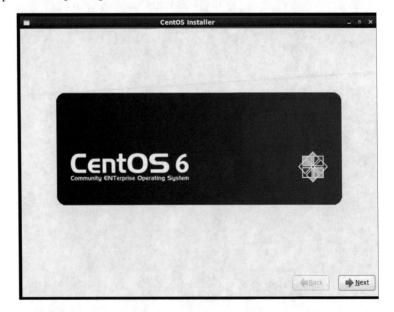

Step 14: Press [Next] in CentOS system to enter the installation setting menu.

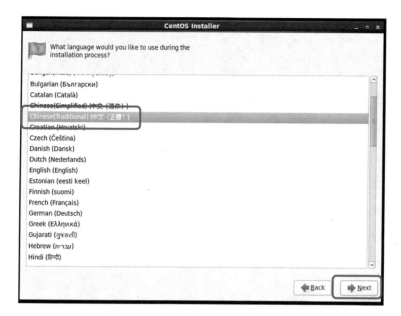

Step 15: Setting the language of installation, in this installation using Chinese.

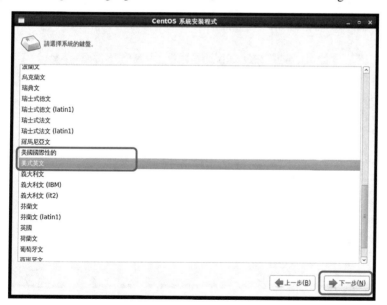

Step 16: Setting the language of installation.

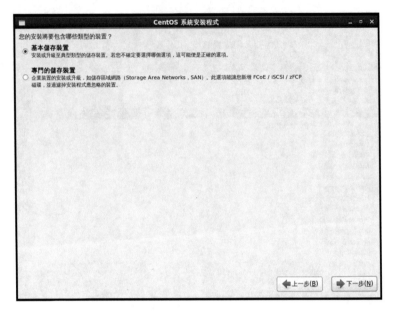

Step 17: Select [Basic Storage Device].

Step 18: Select [Abandon Data].

Step 19: Name the host computer.

Step 20: Select region (Taipei or USA), then go next setp.

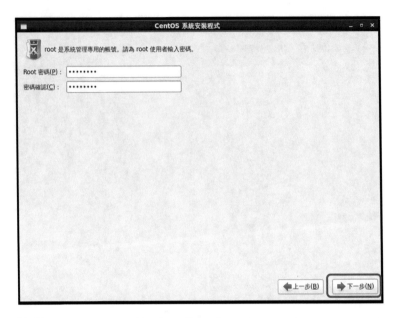

Step 21: Enter root password and confirm, then go next setp.

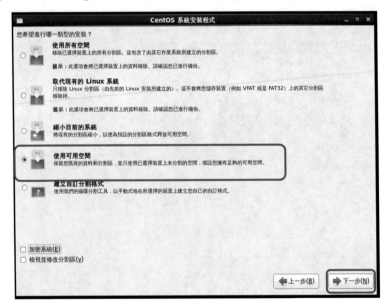

Step 22: Select [Use the available space] in the next step just to be safe, and then go next setp.

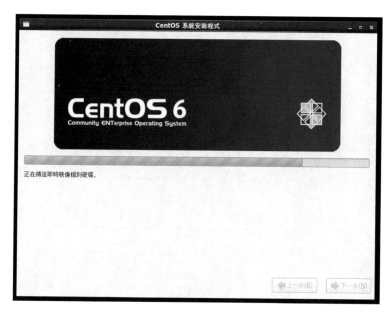

Step 23: Start the installation of CentOS.

Step 24: SET F to restart CentOS once after successful installation.

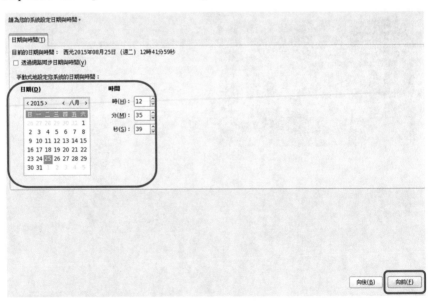

Step 25: Enter F [Name and Password].

Step 26: [Select Time]. This time must be consistent with the actual time of the moment otherwise there might be malfunction. Or the synchronization via the Internet can also be selected.

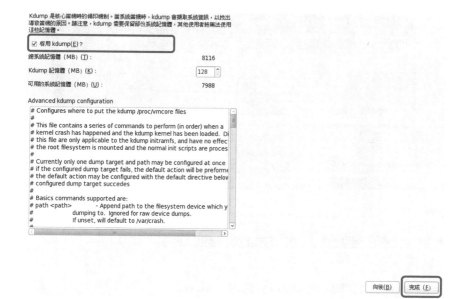

Step 27: Click select [Start kdump(E)] and then push complete bottom.

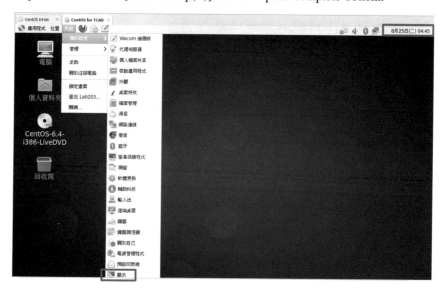

Step 28: The [Display] on the System Preference Menu can be selected to change the screen resolution. The time on the upper right corner of the screen must be changed to the local time, or TCAD will not function normally.

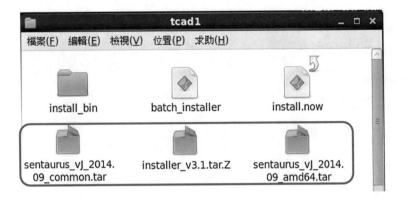

3. Synopsys Sentaurus TCAD software installation

Step 29: Install all files required by TCAD_2014:

(1) **installer_v3.1.tar.Z**
(2) **sentaurus_vj_2014.09_common.tar**
(3) **sentaurus_vj_2014.09_amd64.tar**

All aforementioned files can be accessed via FTP connection to (1) Synopsys; (2) National Center of High-Performance Computing of all countries. For example: National Center for High-performance Computing (NCHC) of Taiwan at https://www.nchc.org.tw/tw/, where three files will be saved in the same folder.

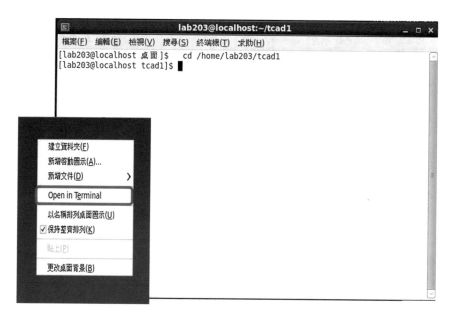

Step 30: Right click the desktop and select [open in Terminal] to access the terminal and type [cd /home/lab203/tcad1] (© Synopsys, InC).

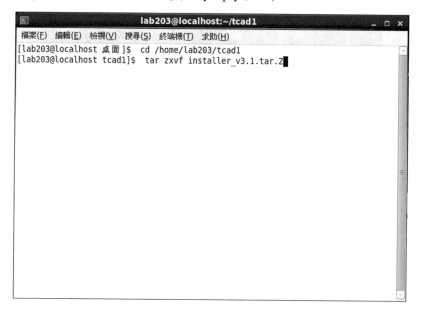

Step 31: Type [tar zxcv installer_v3.1.tar.Z] for decompression.

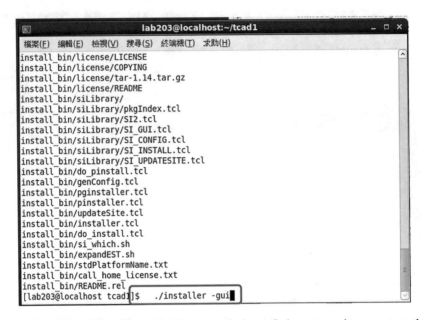

Step 32: Type [./installer-gui] after completion of decompression to start the installation of graphic interface.

Step 33: Press [Start] to start the installation.

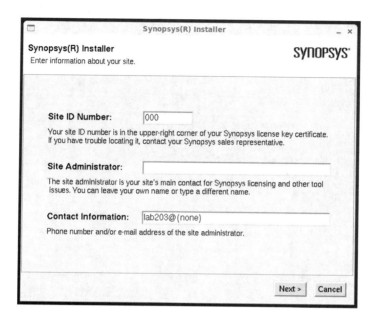

Step 34: Press [Next] to keep the default setting.

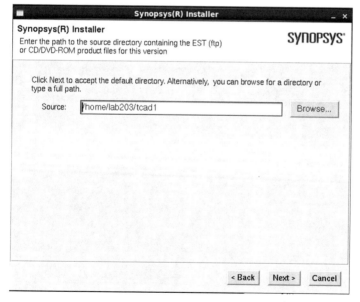

Step 35: Press [Next] to keep the default setting.

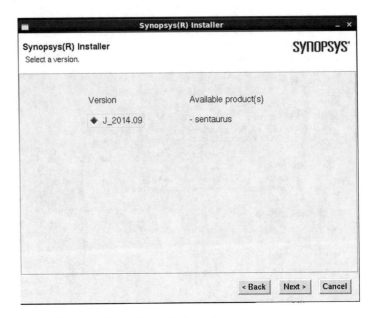

Step 36: Press [Next] to keep the default setting.

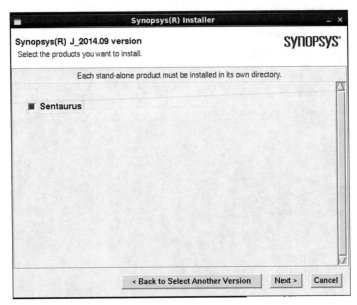

Step 37: Press [Next] to keep the default setting.

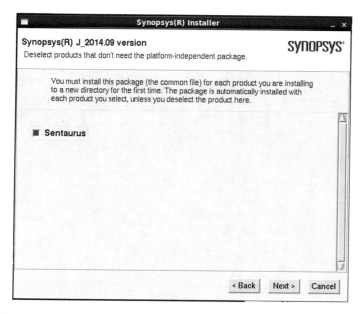

Step 38: Press [Next] to keep the default setting.

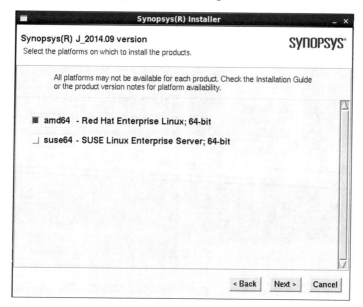

Step 39: Select Red Hat or SUSE in accordance with the version of Linux before pressing [Next].

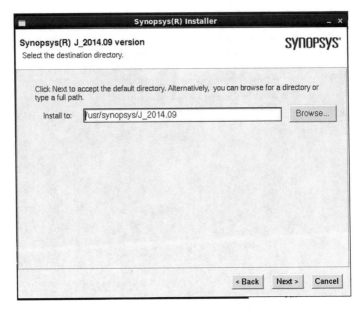

Step 40: This step should be changed to home to avoid any problem.

Step 41: The next step is for setting environmental variables. The first thing is to switch to root account.

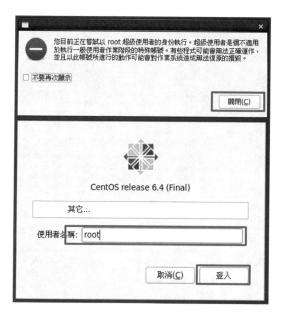

Step 42: Enter root as the ID and password and press Login, and then press Close on the popped up window.

Step 43: Press Search File.

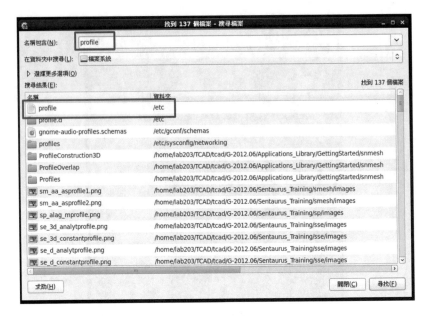

Step 44: Search for [profile] and then open this file.

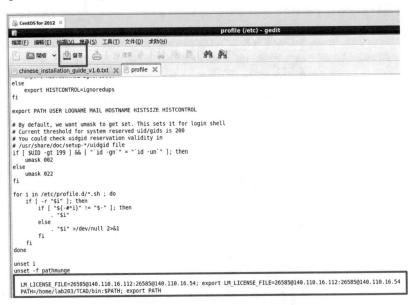

Step 45: The two lines at the bottom should be added. The path for TCAD installation should be entered as the path, and those following/bin will not be changed. Press Save on the top bar after this step is completed.

Step 46: Leave the root account and return to the original account. Access the terminal and **type source/etc/profile**.

Step 47: And then type export |grep LM_LICENSE_FILE and export |grep PATH to verify the correctness of environmental variables.

Step 48: **Enter swb& to start the program**.

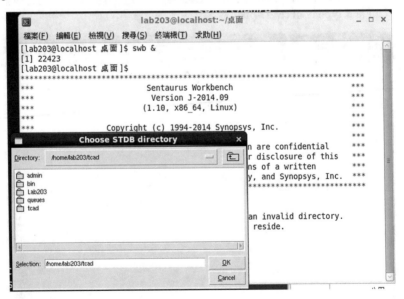

Step 49: Select proper location for saved file, and it can be a self-defined folder. And then press OK to start the simulation program.

Printed in the United States
By Bookmasters